Lecture Notes in Economics and Mathematical Systems

(Vol. 1–15: Lecture Notes in Operations Research and Mathematical Economics, Vol. 16–59: Lecture Notes in Operations Research and Mathematical Systems) For information about Vols. 1–29, please contact your bookseller or Springer-Verlag

Vol. 30: H. Noltemeier, Sensitivitätsanalyse bei diskreten linearen Optimierungsproblemen. VI, 102 Seiten. 1970.

Vol. 31: M. Kühlmeyer, Die nichtzentrale t-Verteilung. II, 106 Seiten. 1970.

Vol. 32: F. Bartholomes und G. Hotz, Homomorphismen und Reduktionen linearer Sprachen. XII, 143 Seiten. 1970. DM 18,–

Vol. 33: K. Hinderer, Foundations of Non-stationary Dynamic Programming with Discrete Time Parameter. VI, 160 pages. 1970.

Vol. 34: H. Störmer, Semi-Markoff-Prozesse mit endlich vielen Zuständen. Theorie und Anwendungen. VII, 128 Seiten. 1970.

Vol. 35: F. Ferschl, Markovketten. VI, 168 Seiten. 1970.

Vol. 36: M. J. P. Magill, On a General Economic Theory of Motion. VI, 95 pages. 1970.

Vol. 37: H. Müller-Merbach, On Round-Off Errors in Linear Programming. V, 48 pages. 1970.

Vol. 38: Statistische Methoden I. Herausgegeben von E. Walter. VIII, 338 Seiten. 1970.

Vol. 39: Statistische Methoden II. Herausgegeben von E. Walter. IV, 157 Seiten. 1970.

Vol. 40: H. Drygas, The Coordinate-Free Approach to Gauss-Markov Estimation. VIII, 113 pages. 1970.

Vol. 41: U. Ueing, Zwei Lösungsmethoden für nichtkonvexe Programmierungsprobleme. IV, 92 Seiten. 1971.

Vol. 42: A. V. Balakrishnan, Introduction to Optimization Theory in a Hilbert Space. IV, 153 pages. 1971.

Vol. 43: J. A. Morales, Bayesian Full Information Structural Analysis. VI, 154 pages. 1971.

Vol. 44: G. Feichtinger, Stochastische Modelle demographischer Prozesse. IX, 404 Seiten. 1971.

Vol. 45: K. Wendler, Hauptaustauschschritte (Principal Pivoting). II, 64 Seiten. 1971.

Vol. 46: C. Boucher, Leçons sur la théorie des automates mathématiques. VIII, 193 pages. 1971.

Vol. 47: H. A. Nour Eldin, Optimierung linearer Regelsysteme mit quadratischer Zielfunktion. VIII, 163 Seiten. 1971.

Vol. 48: M. Constam, FORTRAN für Anfänger. 2. Auflage. VI, 148 Seiten. 1973.

Vol. 49: Ch. Schneeweiß, Regelungstechnische stochastische Optimierungsverfahren. XI, 254 Seiten. 1971.

Vol. 50: Unternehmensforschung Heute – Übersichtsvorträge der Züricher Tagung von SVOR und DGU, September 1970. Herausgegeben von M. Beckmann. IV, 133 Seiten. 1971.

Vol. 51: Digitale Simulation. Herausgegeben von K. Bauknecht und W. Nef. IV, 207 Seiten. 1971.

Vol. 52: Invariant Imbedding. Proceedings 1970. Edited by R. E. Bellman and E. D. Denman. IV, 148 pages. 1971.

Vol. 53: J. Rosenmüller, Kooperative Spiele und Märkte. III, 152 Seiten. 1971.

Vol. 54: C. C. von Weizsäcker, Steady State Capital Theory. III, 102 pages. 1971.

Vol. 55: P. A. V. B. Swamy, Statistical Inference in Random Coefficient Regression Models. VIII, 209 pages. 1971.

Vol. 56: Mohamed A. El-Hodiri, Constrained Extrema. Introduction to the Differentiable Case with Economic Applications. III, 130 pages. 1971.

Vol. 57: E. Freund, Zeitvariable Mehrgrößensysteme. VIII,160 Seiten. 1971.

Vol. 58: P. B. Hagelschuer, Theorie der linearen Dekomposition. VII, 191 Seiten. 1971.

Vol. 59: J. A Hanson, Growth in Open Economies. V, 128 pages. 1971.

Vol. 60: H. Hauptmann, Schätz- und Kontrolltheorie in stetigen dynamischen Wirtschaftsmodellen. V, 104 Seiten. 1971.

Vol. 61: K. H. F. Meyer, Wartesysteme mit variabler Bearbeitungsrate. VII, 314 Seiten. 1971.

Vol. 62: W. Krelle u. G. Gabisch unter Mitarbeit von J. Burgermeister, Wachstumstheorie. VII, 223 Seiten. 1972.

Vol. 63: J. Kohlas, Monte Carlo Simulation im Operations Research VI, 162 Seiten. 1972.

Vol. 64: P. Gessner u K. Spremann, Optimierung in Funktionenräumen. IV, 120 Seiten. 1972

Vol. 65: W. Everling, Exercises in Computer Systems Analysis. VIII, 184 pages. 1972.

Vol. 66: F. Bauer, P. Garabedian and D. Korn, Supercritical Wing Sections. V, 211 pages. 1972.

Vol. 67: I. V. Girsanov, Lectures on Mathematical Theory of Extremum Problems. V, 136 pages. 1972.

Vol. 68: J. Loeckx, Computability and Decidability. An Introduction for Students of Computer Science. VI, 76 pages. 1972.

Vol. 69: S. Ashour, Sequencing Theory. V, 133 pages. 1972.

Vol. 70: J. P. Brown, The Economic Effects of Floods. Investigations of a Stochastic Model of Rational Investment. Behavior in the Face of Floods. V, 87 pages. 1972.

Vol. 71: R. Henn und O. Opitz, Konsum- und Produktionstheorie II V, 134 Seiten. 1972.

Vol. 72: T. P. Bagchi and J. G. C. Templeton, Numerical Methods in Markov Chains and Bulk Queues. XI, 89 pages. 1972.

Vol. 73: H. Kiendl, Suboptimale Regler mit abschnittweise linearer Struktur. VI, 146 Seiten. 1972.

Vol. 74: F. Pokropp, Aggregation von Produktionsfunktionen. VI, 107 Seiten. 1972.

Vol. 75: GI-Gesellschaft für Informatik e.V. Bericht Nr. 3. 1. Fachtagung über Programmiersprachen · München, 9.–11. März 1971. Herausgegeben im Auftrag der Gesellschaft fur Informatik von H. Langmaack und M. Paul. VII, 280 Seiten 1972.

Vol. 76: G. Fandel, Optimale Entscheidung bei mehrfacher Zielsetzung. II, 121 Seiten. 1972.

Vol. 77: A. Auslender, Problèmes de Minimax via l'Analyse Convexe et les Inégalités Variationelles. Théorie et Algorithmes. VII, 132 pages. 1972.

Vol. 78: GI-Gesellschaft für Informatik e.V. 2. Jahrestagung, Karlsruhe, 2.–4. Oktober 1972. Herausgegeben im Auftrag der Gesellschaft für Informatik von P. Deussen. XI, 576 Seiten. 1973.

Vol. 79: A. Berman, Cones, Matrices and Mathematical Programming. V, 96 pages. 1973.

Vol. 80: International Seminar on Trends in Mathematical Modelling, Venice, 13–18 December 1971. Edited by N. Hawkes. VI, 288 pages. 1973.

Vol. 81: Advanced Course on Software Engineering. Edited by F. L. Bauer. XII, 545 pages. 1973.

Vol. 82: R. Saeks, Resolution Space, Operators and Systems. X, 267 pages. 1973.

continuation on page 201

Lecture Notes in Economics and Mathematical Systems

Managing Editors: M. Beckmann and H. P. Künzi

Econometrics

192

Herman J. Bierens

Robust Methods and Asymptotic Theory in Nonlinear Econometrics

Springer-Verlag
Berlin Heidelberg New York 1981

Author

Herman J. Bierens
Foundation for Economic Research of the University of Amsterdam
Jodenbreestraat 23, 1011NH Amsterdam, The Netherlands

AMS Subject Classifications (1970): 62P20

ISBN-13: 978-3-540-10838-2 e-ISBN-13: 978-3-642-45529-2
DOI: 10.1007/978-3-642-45529-2

2142/3140-543210

Voor CHERIE, LAURENS en ROWIN

PREFACE

This Lecture Note deals with asymptotic properties, i.e. weak and strong consistency and asymptotic normality, of parameter estimators of nonlinear regression models and nonlinear structural equations under various assumptions on the distribution of the data. The estimation methods involved are nonlinear least squares estimation (NLLSE), nonlinear robust M-estimation (NLRME) and non-linear weighted robust M-estimation (NLWRME) for the regression case and nonlinear two-stage least squares estimation (NL2SLSE) and a new method called minimum information estimation (MIE) for the case of structural equations.
The asymptotic properties of the NLLSE and the two robust M-estimation methods are derived from further elaborations of results of Jennrich. Special attention is payed to the comparison of the asymptotic efficiency of NLLSE and NLRME. It is shown that if the tails of the error distribution are fatter than those of the normal distribution NLRME is more efficient than NLLSE. The NLWRME method is appropriate if the distributions of both the errors and the regressors have fat tails.
This study also improves and extends the NL2SLSE theory of Amemiya. The method involved is a variant of the instrumental variables method, requiring at least as many instrumental variables as parameters to be estimated. The new MIE method requires less instrumental variables. Asymptotic normality can be derived by employing only one instrumental variable and consistency can even be proved with-out using any instrumental variables at all.
The asymptotic results are not only derived under the assumption that the observations are independently distributed but also for the case where some explanatory variables are lagged dependent variables.
The last chapter deals with some empirical applications of the NLRME method and the MIE method.

This study is a slightly revised version of my Ph.D. dissertation submitted to the Faculty of Economics of the University of Amsterdam. I would like to express my thanks to professor Christopher A. Sims of the University of Minnesota for his willingness to act as my Ph. D. supervisor. His helpful suggestions and remarks benefitted the final results and stimulated me to go on with the lonely job of writing this book. Professor J. Th. Runnenburg guided me on the slippery path of probability theory. Being a self-taught man in this matter, I am much indebted to him for his criticism on previous drafts and for showing me how to do the mathematics properly. Of course, only I am responsible for any remaining errors or sloppiness. I also wish to express my thanks to my teacher professor J.S. Cramer who read the manuscript and offered valuable suggestions.

Preliminary versions of some parts of this book have been disseminated as working papers. I acknowledge the comments of professor Robert I. Jennrich who made we aware of the literature on robust M-estimation and professor Benoit Mandelbrot who suggested some additional references.

I have written this book during the time I was a research fellow of the Foundation for Economic Research of the University of Amsterdam. The Foundation provided me with time, computer facilities and secretarial help. I especially express my gratitude to Mrs. Tjiam Hoa Nio for typing the various drafts.

April 1981 Herman J. Bierens

CONTENTS Page

1 INTRODUCTION

1.1 Specification and misspecification of the econometric model

Econometrics can be characterized as the science which is dealing with obtaining information from economic data, using mathematical-statistical methods. This description covers theoretical econometrics which is concerned with the question what the best way is of obtaining this information, as well as empirical econometrics which is concerned with doing the actual job. Application of mathematical-statistical methods implies the use of econometric models. An econometric model is [Cramer (1969, p.3-4)] "a set of hypotheses that permits statistical inference from the particular data under view". These hypotheses are borrowed partly from economic theory. However, [Cramer (1969, p.2)]" unfortunately economic theorists set great store by generality, and their models are therefore as a rule insufficiently specific to permit an empirical application. As a consequence, virtually all econometric studies add specific hypotheses of their own which are appropiate to the particular situation under review. These convenient approximations are dictated by the requirements of statistical estimation; they are based on common sense rather than on abstract economic theory". Of course not all economic theory is badly specified, especially consumption theory where the form of the demand function is derived by maximizing utility under budget constrains. But even then the utility function has to be specified.

Even when the functional form of the model is supplied by economic theory, the econometrician has to make assumptions about the stochastic nature of the model because economic theory is usually deterministic while deterministic economic models never fit the data perfectly. Thus a disturbance*) term is added to the model and usually the assumption is made that this disturbance term has zero expectation and finite variance or even that it is normally distributed.

The assumption of normality of the disturbance term is usually based on the argument that this disturbance term represents the total impact of a large number of variables not considered in the model, and each having only a small impact itself. These small impacts are considered as independent random drawings from a distribution with finite first and second moments and then, refering to the central limit theorem, normality is postulated. However, why should economic

---

*) Which will also be called "error term" or shortly "error".

**) Following the Dutch convention, random variables, - vectors and - functions are underlined, to distinguish them from values they may take.

variables have finite moments at all? It is well known that the distribution of the highest incomes can be described by the Pareto law, that is: if $\underline{y}$**) is a random drawing from the population of incomes larger than y_0, then $P(\underline{y} \leq y) = 1-(\frac{y}{y_0})^{\alpha}$, $\alpha > 0$ provided that y_0 is large enough. This distribution has a finite second moment only if $\alpha > 2$, while in empirical research α's are found which are smaller than 2. Also other economic variables may have distributions without finite second moment. Mandelbrot (1963 a,b, 1967) and Fama(1963, 1965) have found empirical as well as theoretical evidence for such distributions, especially with respect to speculative prices. But if some economic variables have infinite second moments, then the same may be the case for disturbance terms. Not recognizing this may have serious consequences for least squares estimators. For, if $\underline{y}_j = \beta x_j + \underline{u}_j$, $j=1,2,\ldots$, where the x_j's are non-stochastic explanatory variables and the $\underline{u}_j$'s are random-drawings from the Cauchy $(0,\sigma)$ distribution [N.B.: the density of this Cauchy distribution is $f(u) = \frac{1}{\pi}\frac{\sigma}{\sigma^2+u^2}$] then the least-squares estimator $\underline{b}_n$, say, of β is also Cauchy distributed:

$$\underline{b}_n = \frac{\sum_{j=1}^n x_j \underline{y}_j}{\sum_{j=1}^n x_j^2} = \beta + \frac{\sum_{j=1}^n x_j \underline{u}_j}{\sum_{j=1}^n x_j^2} \sim \text{Cauchy}\left(\beta, \sigma . \frac{\sum_{j=1}^n |x_j|}{\sum_{j=1}^n x_j^2}\right) .$$

Moreover if $c_1 = \lim \frac{1}{n}\sum_{j=1}^n |x_j|$ and $c_2 = \lim \frac{1}{n}\sum_{j=1}^n x_j^2$ both exist and are positive then $\underline{b}_n$ converges in distribution to Cauchy $(\beta, \sigma . \frac{c_1}{c_2})$, so that the least squares estimator $\underline{b}_n$ fails to be consistent. In the case that the $\underline{u}_j$'s are i.i.d. $N(0,\sigma^2)$ we obviously have plim $\underline{b}_n = \beta$, provided that $\sum_{j=1}^n x_j^2 \to \infty$ as $n \to \infty$.

We have just mentioned, referring to the work of Mandelbrot and Fama, that it is not necessary that distributions of economic variables have finite variances. The argument of these authors is partly based on a theorem from probability theory [see Feller (1966) chapter XVII], which states that if a partial sum $\sum_{j=1}^n \underline{u}_j$ of i.i.d. random variables has a non-degenerate limiting distribution G, say, (which means that there are numbers a_n, b_n such that $\lim_{n\to\infty} P(a_n(\sum_{j=1}^n \underline{u}_j - b_n) \leq x) = G(x)$ for all continuity points x of G), then this distribution function G belongs to the class of stable distributions. These stable distributions can be characterized by some parameters, of which the most important one is the so-called characteristic exponent $\alpha \in (0,2]$.

Among these distributions only the normal (the case $\alpha=2$) has a finite variance.

Another example of a stable distribution is the Cauchy distribution (the case $\alpha=1$), but for other members of this class with characteristic exponent $\alpha\in(0,1)$ or $\alpha\in(1,2)$ no closed form of the distribution function is available yet, even in the symmetric case where the density of a symmetric stable distribution with characteristic exponent α can only be written as

$$f(u)=\frac{1}{2\pi}\int_{-\infty}^{+\infty}\cos(tu)e^{-\gamma|t|^{\alpha}}dt\ ,\ \gamma>0,\quad \alpha\in(0,2]\ ,$$

where γ is a scaling parameter.

Thus the usual argument for underpinning the normality assumption of disturbance terms does not necessarily lead to the conclusion that disturbance terms are normal, but only that a stable distribution is appropriate. Of course, the above argument does not imply that error distributions are necessarily stable, because the impacts of the variables not taken into account may fail to be independent or equally distributed. We are only making a plea to break the automatism of assuming normality of the error distribution, to recognize the possibility that error distributions may have (nearly) infinite variances and to use estimation methods being able to cope with such situations. In connection with this we note that deviations from normality of the errors of a regression model may cause the occurence of outliers of the residuals and that the performance of the least squares estimation method is rather sensitive to outliers.

Specification of the distribution of the disturbance term is only one side of the specification problem. Often economic theory is vague about the functional form of a relationship and in such cases usually the linear regression model is assumed. But misspecification of the functional form may also have serious consequences for the conclusions to be drawn. For example if the relationship between $\underline{y}$ and $\underline{x}$ is $\underline{y}_j=\underline{x}_j^2+\underline{u}_j$, $j=1,2,\ldots$, where the pairs $(\underline{x}_j,\underline{u}_j)$ are independent random drawings from the bivariate normal distribution $N_2\left[\binom{0}{0},\begin{pmatrix}\sigma_x^2 & 0\\ 0 & \sigma_u^2\end{pmatrix}\right]$ and if the model is specified as $\underline{y}_j=\beta\underline{x}_j+\underline{u}_j$, $j=1,2,\ldots$, then the least squares estimator

$$\underline{b}_n=\frac{\sum_{j=1}^{n}\underline{x}_j\underline{y}_j}{\sum_{j=1}^{n}\underline{x}_j^2}=\frac{\frac{1}{n}\sum_{j=1}^{n}\underline{x}_j^3+\frac{1}{n}\sum_{j=1}^{n}\underline{x}_j\,\underline{u}_j}{\frac{1}{n}\sum_{j=1}^{n}\underline{x}_j^2}$$

converges in probability to zero because

$$\operatorname{plim}\frac{1}{n}\sum_{j=1}^{n}\underline{x}_j^3=0,\ \operatorname{plim}\frac{1}{n}\sum_{j=1}^{n}\underline{x}_j\underline{u}_j=0 \text{ and } \operatorname{plim}\frac{1}{n}\sum_{j=1}^{n}\underline{x}_j^2=\sigma_x^2.$$

Thus we would conclude (when n is sufficiently large) that $\beta=0$ and hence that the theory involved is false. Of course, this conclusion is correct as far as the econometric hypotheses are concerned. But usually econometricians then also reject the economic theory involved.

In the case that the functional form of the model is not supplied by economic theory, it is very difficult to avoid misspecification of this functional form because often the number of theoretically possible forms is infinite.From a practical point of view the best way seems be to specify the functional forms as flexibly as possible, for example by using the well known Box-Cox transformation

$$\phi_\lambda(x) = \frac{x^\lambda - 1}{\lambda}, \quad x > 0$$

[Box-Cox (1964)] so that then the functional form is partly determined by the data itself. Instead of specifying a linear model $\underline{y}_j = \sum_{i=1}^k \beta_i \underline{x}_{i,j} + \underline{u}_j$, or a log-linear model $\log(\underline{y}_j) = \sum_{i=1}^k \beta_i \log(\underline{x}_{i,j}) + \underline{u}_j$ we then put the model in the form:

$$\phi_{\lambda_0}(\underline{y}_j) = \sum_{i=1}^k \beta_i \phi_{\lambda_i}(\underline{x}_{i,j}) + \underline{u}_j, \quad j=1,2,\ldots\ldots,$$

where now $\lambda_0, \lambda_1, \ldots \lambda_k, \beta_1, \ldots, \beta_k$ are the parameters to be estimated.

The Box-Cox transformation is applied by Zarembka (1968), White (1972) and Spitzer (1976) to determine the functional form of the demand function of money, while Leech (1975) has used it to determine whether the disturbance term of a CES-production function is additive or multiplicative. In all these cases the maximum likelihood method is used, assuming normality of the error distribution, and hypotheses with respect to λ are tested by the likelihood ratio test [see Goldfeld and Quandt (1972)]. But in view of our previous argument about the error distribution, the normality assumption may be too strong, hence the optimal properties of the maximum likelihood method may not apply at all.

We can specify our functional form always as flexible as we wish, for example by using polynominal transformations, but then we also have to introduce more parameters. Obviously the number of parameters that can be handled is limited by the number of observations, which is usually small in econometrics. So in practice linear specification will often be unavoidable. Therefore we shall deal with classes of nonlinear models that also contain the linear case.

1.2 The purpose and scope of this study

From the previous consideration the conclusion can be drawn that there is a need for estimation methods which are appropriate for nonlinear models as well as robust. Malinvaud (1970, p.296) states: "One describes as robust those statistical procedures, which show little sensitivity to the assumptions on the stochastic variables of the model for which they are conceived". The main purpose of this study is to provide such procedures and to study their asymptotic properties, but we shall also pay attention to the most important existing nonlinear estimation methods, namely nonlinear least squares and nonlinear two-stage least squares.

Although we shall consider error distributions belonging to a more general class than only the normal, it will be necessary to impose some restrictions on this class. Except in the cases of nonlinear least squares and two-stage least squares estimation, we shall mainly consider symmetric error distributions, and dealing with regression models with additive errors we shall moreover assume that the error distribution is unimodal. Since symmetric stable distributions are unimodal [see theorem 2.5.4 in section 2.5 below and the remark of Doob in the appendix of Gnedenko and Kolmogorov (1954)], the symmetric unimodal case covers the symmetric stable case.

For studying the asymptotic properties of nonlinear estimators we need much more probability theory than in the linear case and even some contributions to probability theory itself have to be made, especially concerning convergence of random functions. This preliminary mathematical theory is provided by chapter 2. In chapter 3 we consider the problem of estimating the parameters of nonlinear regression models under various assumptions about the distribution of the errors and the regressors. The asymptotic properties of three types of estimators are studied, namely nonlinear least squares estimators, robust M-estimators and weighted robust M-estimations. In chapter 4 we shall deal with estimation of a single nonlinear implicit structural equation. We first consider the asymptotic properties of the nonlinear two-stage least squares estimator , and then we propose a new estimator which requires less instrumental variables than the former method. In these latter two chapters it is throughout assumed that the observations are independently distributed, but in chapter 5 this assumption is dropped. Instead of independence, a new stability condition for nonlinear auto-regressive stochastic processes is introduced. Under this stability condition the convergence in probability and convergence in distribution results of the chapters 3 and 4 carry over. Finally, in chapter 6 we present some applications.

2 PRELIMINARY MATHEMATICS

For understanding linear econometrics, a good background in calculus, statistics and linear algebra may be sufficient, but for nonlinear econometrics we need additional knowledge of abstract probability theory. Since this book is mainly written for econometricians and not for mathematicians, it is presumed that the reader is not completely familiar with the measure-theoretical approach of probability theory. We shall therefore review and explain this additional mathematics in the sections 2.1 and 2.2, in order to make this study (nearly) self-contained. Most of the material can be found in textbooks like Chung (1974) and Feller (1966), but the theorems 2.2.15 through 2.2.17 are our own elaborations of known results.

Section 2.3 deals with uniform convergence of random functions on compact spaces. Especially the theorems 2.3.3 and 2.3.4, which are further elaborations of results of Jennrich (1969), are very important for us. Section 2.4 gives a brief review of some results on characteristic functions and stable distributions. Moreover, we state there the famous central limit theorem of Liapounov. Finally, section 2.5 is devoted to properties of (symmetric) unimodal distributions.

2.1 <u>Random variables, independence, Borel measurable functions and mathematical expectation</u>

2.1.1 <u>Measure theoretical foundation of probability theory</u>

In this section we shall give a brief outline of the measure-theoretical foundation of probability theory. Dealing with convergence of random variables and uniform convergence of random functions, measure-theoretical arguments are unavoidable. These convergence concepts will play a key role in this study.

The basic concept of probability theory is the <u>probability space</u>. This is a triple $\{\Omega,\mathfrak{F},P\}$ consisting of:

1. an abstract non empty set Ω, called the <u>sample space</u>. We do not impose any conditions on this set.
2. a non empty collection $\mathfrak{F}$ of subsets of Ω, having the following two properties:
   - if $E\in\mathfrak{F}$, then $E^c\in\mathfrak{F}$, (2.1.1)

     where E^c denotes the complement of the subset E with respect to Ω: $E^c = \Omega\setminus E$
   - if $E_j\in\mathfrak{F}$ for $j=1,2,\ldots$, then $\bigcup_{j=1}^{\infty} E_j\in\mathfrak{F}$. (2.1.2)

   These two properties make $\mathfrak{F}$, by definition, a <u>Borelfield</u> of subsets of Ω.

3. a probability measure P on $\{\Omega,\mathcal{F}\}$. This is a real valued set function on $\mathcal{F}$ such that:
   - $P(\Omega)=1$ (2.1.3)
   - $P(E)\geq 0$ for all $E\in\mathcal{F}$ (2.1.4)
   - $E_j\in\mathcal{F}$ for $j=1,2,\ldots$ and $E_{j_1}\cap E_{j_2}=\emptyset$ if $j_1\neq j_2$ imply $P(\bigcup_{j=1}^{\infty})=\sum_{j=1}^{\infty}P(E_j)$. (2.1.5)

   In words: The probability of any union of disjoint sets in $\mathcal{F}$ equals the sum of the probabilities of each of these sets.

Now let $\underline{x}$ *) be a random variable and let F be its distribution function. In the measure-theoretical approach of probability theory a random variable is considered as a real valued function on the set Ω denoted by:

$\underline{x} = x(.)$

with value $x(\omega)$ at $\omega\in\Omega$, such that for every real number t:

$\{\omega\in\Omega : x(\omega)\leq t\}\in\mathcal{F}$.

The distribution function F with value F(t) at $t\in R$ is then defined by

$F(t) = P(\{\omega\in\Omega : x(\omega)\leq t\})$,

which will often be denoted by the short-hand notation:

$F(t) = P(\underline{x}\leq t)$.

It is not hard to see that the axioms (2.1.1) and (2.1.2) imply:

$E_j\in\mathcal{F},j=1,2,\ldots \Rightarrow \bigcap_{j=1}^{\infty}E_j\in\mathcal{F}$

and that from (2.1.3), (2.1.4) and (2.1.5) follow:

$P(\emptyset) = 0$

$P(E^c) = 1 - P(E)$

$P(E\cup F) + P(E\cap F) = P(E) + P(F)$

$E\subset F \Rightarrow P(E) \leq P(F\setminus E)+P(E)= P(F)$

$E_n\subset E_{n+1},\ E = \bigcup_{n=1}^{\infty} E_n \Rightarrow P(E_n) \to P(E)$

$E_n\supset E_{n+1},\ E = \bigcap_{n=1}^{\infty} E_n \Rightarrow P(E_n) \to P(E)$

$P(\bigcup_{j=1}^{\infty} E_j)\leq \sum_{j=1}^{\infty} P(E_j)$,

where all sets involved are members of $\mathcal{F}$. Moreover, the distribution function F(t) is right continuous: $F(t) = \lim_{\varepsilon\downarrow 0} F(t+\varepsilon)$, as is easily verified, and it satisfies

$F(\infty) = \lim_{t\to\infty} F(t) = 1,\ F(-\infty) = \lim_{t\to-\infty} F(t) = 0.$

*) We recall that random variables,- vectors and - functions are underlined.

Furthermore, by F(t-) we denote

$$F(t-) = \lim_{\varepsilon \downarrow 0} F(t-\varepsilon) ,$$

which clearly satisfies $F(t-) \leq F(t)$.

A finite dimensional random vector can now be defined as a vector with random variables as components, where these random components are assumed to be defined on a common probability space.
A complex random variable, $\underline{z} = \underline{x} + i\underline{y}$ is now a complex valued functions

$$\underline{z} = z(.) = x(.) + iy(.)$$

on Ω with real valued random variables $\underline{x}$ and $\underline{y}$, i.e.

$$\{\omega \in \Omega : x(\omega) \leq t\} \in \mathcal{F} , \{\omega \in \Omega : y(\omega) \leq t\} \in \mathcal{F}$$

for any real number t.

Next we shall construct a Borelfield $\mathcal{B}$ of subsets of R^k such that for every set $B \in \mathcal{B}$ and any k-dimensional random vector $\underline{x}$ on a probability space $\{\Omega, \mathcal{F}, P\}$ we have

$$\{\omega \in \Omega : x(\omega) \in B\} \in \mathcal{F} , \tag{2.1.6}$$

because only for such subsets B of R^k we can define the probability

$$P(\underline{x} \in B) = P(\{\omega \in \Omega : x(\omega) \in B\}) , \tag{2.1.7}$$

since probability is only defined for members of the Borelfield $\mathcal{F}$.
Let $\mathcal{C}$ be the collection of subsets of R^k of the type $\times_{m=1}^{k} (-\infty, t_m]$, $t_m \in R$, and let $\mathcal{G}$ be the Borelfield of all subsets of R^k.

Clearly we have $\mathcal{C} \subset \mathcal{G}$, that means that $E \in \mathcal{C} \Rightarrow E \in \mathcal{G}$. But next to $\mathcal{G}$ there may be other Borelfields of subsets of R^k with this property, say $\mathcal{G}_\alpha$, $\alpha \in A$, where A is an index set. Assuming that all Borelfields containing $\mathcal{C}$ are represented this way, we then have a non-empty collection of Borelfields $\mathcal{G}_\alpha$, $\alpha \in A$, of subsets of R^k such that $\mathcal{C} \subset \mathcal{G}_\alpha$ for any $\alpha \in A$. Now consider the collection $\mathcal{B} = \cap_{\alpha \in A} \mathcal{G}_\alpha$. Since each $\mathcal{G}_\alpha$ is a Borelfield, it follows that this collection $\mathcal{B}$ is a Borelfield of subsets of R^k and since $\mathcal{C}$ is contained in each $\mathcal{G}_\alpha$ it follows that $\mathcal{C} \subset \mathcal{B}$. We shall say that the Borelfield $\mathcal{B}$ is the minimal Borelfield containing the collection $\mathcal{C}$, and for this particular collection $\mathcal{C}$ it is called the Euclidean Borelfield. Summarizing:

> Definition 2.1.1. Let $\mathcal{C}$ be any collection of subsets of a set Γ and let the Borelfields of subsets of Γ containing $\mathcal{C}$ be $\mathcal{G}_\alpha$, $\alpha \in A$.
> Then $\mathcal{G} = \cap_{\alpha \in A} \mathcal{G}_\alpha$ is called the minimal Borelfield containing $\mathcal{C}$.

Definition 2.1.2. Let $\mathcal{C}$ be the collection of subsets of R^k of the type $\times_{m=1}^{k}(-\infty,t_m]$, $t_m \in R$. The minimal Borelfield $\mathcal{B}$ containing this collection is called the Euclidean Borelfield and the members of $\mathcal{B}$ are called the Borel sets.

We show that for any Borel set B and any r.v. $\underline{x}$ on $\{\Omega,\mathcal{F},P\}$, (2.1.6) is satisfied. Let $\mathcal{D}$ be the collection of all Borel sets B such that (2.1.6) is satisfied. Then $\mathcal{D} \subset \mathcal{B}$. If $\mathcal{D}$ is also a Borelfield then $\mathcal{B} \subset \mathcal{D}$ (and hence $\mathcal{B} = \mathcal{D}$) because $\mathcal{B}$ is the minimal Borelfield containing the sets $\times_{m=1}^{k}(-\infty,t_m]$ while obviously the collection $\mathcal{D}$ contains these sets. So it suffices to prove that $\mathcal{D}$ is a Borelfield. But this is not hard to do and therefore left to the reader.

Theorem 2.1.1. For any random variable or vector $\underline{x}$ on $\{\Omega,\mathcal{F},P\}$ and any Borel set B we have $\{\omega\in\Omega : x(\omega)\in B\}\in\mathcal{F}$.

Consequently the definition (2.1.7) is meaningful for Borel sets. In fact, by defining a measure μ on the Euclidean Borelfield $\mathcal{B}$ as

$$\mu(B) = P(\underline{x} \in B) = P(\{\omega\in\Omega : x(\omega) \in B\})$$

for any Borel set B we have created a probability measure on $\mathcal{B}$. This probability measure μ is often refered to as the probability measure induced by (the random variable or -vector) $\underline{x}$.

We are now able to define a (joint) distribution function on R^k. Let $\underline{x}$ be a random vector in R^k defined on a probability space $\{\Omega,\mathcal{F},P\}$. The product sets $\times_{j=1}^{k}(-\infty,t_j]$ are Borel sets, where the t_j's are the components of a vector $t\in R^k$.

Thus:

$$\{\omega\in\Omega : x(\omega)\in \times_{j=1}^{k}(-\infty,t_j]\}\in\mathcal{F}.$$

The (joint) distribution function F, say, of $\underline{x}$ is now defined for all $t\in R^k$ by:

$$F(t)= P(\{\omega\in\Omega : x(\omega)\in \times_{j=1}^{k}(-\infty,t_j]\}),$$

which, however, will also often be denoted by the shorthand notation:

$$F(t)= P(\underline{x} \le t).$$

2.1.2 Independence

Now let $\underline{x}_1, \underline{x}_2, \ldots$ be a sequence of random variables with corresponding probability spaces $\{\Omega_1, \mathcal{F}_1, P_1\}, \{\Omega_2, \mathcal{F}_2, P_2\}, \ldots$ respectively. Then it is possible to construct a new probability space, $\{\Omega, \mathcal{F}, P\}$, say, such that the $\underline{x}_j$'s can be regarded as independent random variables on $\{\Omega, \mathcal{F}, P\}$ [see Chung (1974, section 3.3)]. Independence means:

Definition 2.1.3. Let $\underline{x}_1, \underline{x}_2, \ldots$ be random vectors in $R^{k_1}, R^{k_2}, \ldots$, respectively. This sequence $(\underline{x}_j)_{j=1}^{\infty}$ is called independent if for any finite sequence (B_j) of Borel sets (with B_j a Borel set in R^{k_j})

$$P(\bigcap_{j=1}^{m} \{\underline{x}_j \in B_j\}) = \prod_{j=1}^{m} P(\underline{x}_j \in B_j) \quad \text{for all } m \geq 1, \tag{2.1.8}$$

and $(\underline{x}_j)_{j=1}^{\infty}$ is called mutually (or pairwise) independent if for $j_1 \neq j_2$, $\underline{x}_{j_1}$ and $\underline{x}_{j_2}$ are independent.

2.1.3 Borel measurable functions

If $\underline{x}$ is a r.v. and $f(x)$ is a real function on R^1, is then $f(\underline{x})$ a r.v.? The answer is: not always. There are functions [see for example Royden (1968, problem 3.28)] for which this is not the case. The condition for $f(\underline{x})$ being a r.v. is that for all $t \in R^1$ we have $\{\omega \in \Omega : f(x(\omega)) \leq t\} \in \mathcal{F}$, where $\{\Omega, \mathcal{F}, P\}$ is the probability space involved, and refering to theorem 2.1.1 this will be the case if for every $t \in R^1$ the set $\{x \in R^1: f(x) \leq t\}$ is a Borel set. Functions satisfying the latter condition are called Borel measurable. Now consider a real function $f(x_1, \ldots x_k)$ on R^k and r.v.'s $\underline{x}_1, \ldots, \underline{x}_k$ on $\{\Omega, \mathcal{F}, P\}$. If for every $t \in R^1$ the set $\{(x_1, \ldots, x_k) \in R^k: f(x_1, \ldots, x_k) \leq t\}$ is a Borel set B_t, say, in R^k then $\{\omega \in \Omega : f(x_1(\omega), \ldots, x_k(\omega)) \leq t\} \in \mathcal{F}$ for every $t \in R^1$ and hence $f(\underline{x}_1, \ldots, \underline{x}_k)$ is a r.v. Also such functions are called Borel measurable:

Definition 2.1.4. A real function $f(x_1, \ldots, x_k)$ on R^k is called Borel measurable if for every $t \in R^1$ the set $\{(x_1, \ldots, x_k) \in R^k : f(x_1, \ldots, x_k) \leq t\}$ is a Borel set in R^k.

A first example of a Borel measurable function is the so called simple function:

Definition 2.1.5. A real function $f(x_1, \ldots, x_k)$ on R^k is called a simple function if there are finite real numbers $b_1, \ldots, b_n$ and Borel sets $B_j, j=1,2,\ldots,n$ with $B_{j_1} \cap B_{j_2} = \emptyset$ if $j_1 \neq j_2$, such that

$$f(x_1, \ldots, x_k) = \sum_{j=1}^{n} b_j \chi_j(x_1, \ldots, x_k)$$

where

$$\chi_j(x_1, \ldots, x_k) = \begin{cases} 1 & \text{if } (x_1, \ldots, x_k) \in B_j \\ 0 & \text{if } (x_1, \ldots, x_k) \notin B_j \end{cases}$$

If we realize that the set $\{x \in R^k : f(x) \leq t\}$ for a simple function f is always a finite union of Borel sets, we have:

Theorem 2.1.2. Simple functions are Borel measurable.

From this result we can derive other Borel measurable functions using the following theorem.

Theorem 2.1.3. Let $f_1, f_2, \ldots$ be a sequence of Borel measurable functions on R^k. Then the functions $\sup\{f_1, \ldots, f_n\}$, $\inf\{f_1, \ldots, f_n\}$, $\sup_n f_n$, $\inf_n f_n$, $\limsup_n f_n$ and $\liminf_n f_n$ are also Borel measurable.

Proof: We only consider the case k=1. Moreover, it is not hard to see that if f is Borel measurable then -f is Borel measurable, hence it suffices to prove the theorem for the "sup"-cases. Let $h_n(x) = \sup\{f_1(x) \ldots, f_n(x)\}$. Then

$$\{x \in R^1 : h_n(x) \leq t\} = \bigcap_{j=1}^{n} \{x \in R^1 : f_j(x) \leq t\},$$

which is a Borel set since the f_j's are Borel measurable. Moreover, replacing n with ∞ we see that $\sup_n f_n(x)$ is Borel measurable. Since $\limsup_n f_n(x) = \inf_n \sup_{k \geq n} f_k(x)$ and since $\inf_n g_n(x)$ is Borel measurable if each g_n is, it follows directly that $\limsup_n f_n(x)$ is Borel measurable. □

Remark. From this theorem we can also conclude that if $\underline{x}_1, \underline{x}_2, \ldots, \underline{x}_n, \ldots\ldots$ are random variables, so are $\sup\{\underline{x}_1, \ldots, \underline{x}_n\}$, $\inf\{\underline{x}_1, \ldots, \underline{x}_n\}$, $\sup_n \underline{x}_n$, $\inf_n \underline{x}_n$, $\limsup_n \underline{x}_n$ and $\liminf_n \underline{x}_n$.

From the theorems 2.1.2 and 2.1.3 it follows:

Theorem 2.1.4. A continuous real function on R^k is Borel measurable.

Proof: Let f(x) be a continuous real function on R^k. It is not hard to show that the functions $f_n(x)$ defined by

$$f_n(x) = \begin{cases} f(x) & \text{if } |x| \le n \\ 0 & \text{if } |x| > n \end{cases}, \quad n = 1,2, \ldots$$

can, for fixed n, be written as limits of simple functions, so that by theorem 2.1.3 the $f_n(x)$'s are Borel measurable. Since

$$f(x) = \lim_n f_n(x) = \limsup_n f_n(x) = \liminf_n f_n(x)$$

it follows from theorem 2.1.3 that f is Borel measurable. □

Let f be any Borel measurable function on R^k. Since the functions max {0,x} and max {0,-x} are continuous and hence Borel measurable, it follows that

$$f^+(x_1,\ldots,x_k) = \max\{0,f(x_1,\ldots,x_k)\}, f^-(x_1,\ldots,x_k) = \max\{0,-f(x_1,\ldots,x_k)\}.$$

are non-negative Borel measurable functions. Moreover we obviously have

$$f(x_1,\ldots,x_k) = f^+(x_1,\ldots,x_k) - f^-(x_1,\ldots,x_k) \qquad (2.1.9)$$

This representation is important because it means that we can often focus our attention on non-negative Borel measurable functions without loss of generality. Thus the following theorem gives a full characterization of Borel measurable functions:

Theorem 2.1.5. A non-negative real function f on R^k is Borel measurable if and only if there is a non-decreasing sequence of simple functions ψ_n on R^k satisfying $0 \le \psi_n(x_1, \ldots, x_k) \le f(x_1, \ldots, x_k)$ and $\lim_{n\to\infty} \psi_n(x_1, \ldots x_k) = f(x_1, \ldots, x_k)$ for each $(x_1, \ldots, x_k) \in R^k$.

Proof: Take for given non-negative Borel measurable f

$$\psi_n(x_1,\ldots,x_k) = \begin{cases} \frac{m-1}{2^n} & \text{if } \frac{m-1}{2^n} \le f(x_1,\ldots,x_k) \le \frac{m}{2^n} \\ n & \text{otherwise} \end{cases}$$

for integers m with $1 \le m \le n\, 2^n$. Then the ψ_n's have all the required properties. Since by theorem 2.1.2 the simple functions ψ_n are Borel measurable, the limit is Borel measurable by theorem 2.1.3. □

2.1.4 Mathematical expectation

The above theorem can be used for defining the mathematical expectation of $f(\underline{x}_1,\ldots, \underline{x}_k)$, where f is a Borel measurable function on R^k and $\underline{x}_1, \ldots, \underline{x}_k$ are r.v.'s on a probability space $\{\Omega,\mathcal{F},P\}$, as some limit of mathematical expectations of simple functions. The latter expectations are defined as follows:

Definition 2.1.6. Let $f(x_1, \ldots, x_k)$ be the simple function on R^k as defined in definition 2.1.5. and let $\underline{x}_1, \ldots, \underline{x}_k$ be r.v.'s on a probability space $\{\Omega,\mathcal{F},P\}$. Then the mathematical expectation of $f(\underline{x}_1, \ldots, \underline{x}_k)$ is defined by:

$$E\, f(\underline{x}_1, \ldots, \underline{x}_k) = \sum_{j=1}^{n} b_j\ P(\{\omega\in\Omega: (x_1(\omega),\ldots, x_k(\omega))\in B_j\})$$

$$= \sum_{j=1}^{n} b_j\ P(\ \chi_j(\underline{x}_1, \ldots, \underline{x}_k) = 1)$$

For any non-negative Borel measurable function we define:

Definition 2.1.7. Let $f(x_1, \ldots, x_k)$ be a non-negative Borel measurable function on R^k and let $\underline{x}_1, \ldots, \underline{x}_k$ be r.v.'s on a common probability space. Then:

$E\, f(\underline{x}_1, \ldots, \underline{x}_k) = \sup E\, \psi(\underline{x}_1, \ldots, \underline{x}_k)$

where the supremum is taken over all the simple functions satisfying:

$0 \leq \psi(x_1, \ldots, x_k) \leq f(x_1,\ldots,x_k)$ for all $(x_1,\ldots,x_k)\in R^k$.

Using the representation (2.1.9) we now have:

Definition 2.1.8. Let $f(x_1, \ldots, x_k)$ be a Borel measurable function on R^k and let $\underline{x}_1, \ldots, \underline{x}_k$ be r.v. on a common probability space $\{\Omega,\mathcal{F},P\}$.

If $E\, f^+(\underline{x}_1, \ldots, \underline{x}_k) < \infty$ and /or $E\, f^-(\underline{x}_1, \ldots, \underline{x}_k) < \infty$ then:

$$E\, f(\underline{x}_1, \ldots, \underline{x}_k) = E\, f^+(\underline{x}_1, \ldots, \underline{x}_k) - E\, f^-(\underline{x}_1, \ldots, \underline{x}_k)$$

which is also denoted by the general integral with respect to the measure P:

$$E\, f(\underline{x}_1, \ldots, \underline{x}_k) = \int_\Omega f(x_1(\omega), \ldots, x_k(\omega))P(d\omega) = \int_\Omega f(x_1, \ldots, x_k)dP$$

(the last integral is only a short-hand notation of the first one).

If both $E\, f^+(\underline{x}_1, \ldots, \underline{x}_k) = \infty$ and $E\, f^-(\underline{x}_1, \ldots, \underline{x}_k) = \infty$,

the mathematical expectation and the corresponding integral are undefined. Finally, for any set $\Lambda \in \mathcal{F}$ we define

$$\int_{\Lambda} f(x_1(\omega),\ldots,x_k(\omega))P(d\omega)=\int_{\Omega} y(\omega)P(d\omega)$$

where :

$$y(\omega) = \begin{cases} f(x_1(\omega), \ \ldots, \ x_k(\omega)) & \text{if } \omega \in \Lambda \\ 0 & \text{if } \omega \in \Omega \setminus \Lambda, \end{cases}$$

provided that the latter integral is defined.

The mathematical expectation is often denoted by the classical Riemann-Stieltjes integral [see Rudin (1976)]:

$$E\ f(\underline{x}_1, \ldots, \underline{x}_k) = \int_{R^k} f(x_1, \ldots, x_k)dF(x_1,\ldots,x_k), \qquad (2.1.10)$$

where F is the joint distribution function of the random vector $\underline{x}'=(\underline{x}_1,\ldots,\underline{x}_k)$. If the integral (2.1.10) exists for a Borel function f then it equals the mathematical expectation because then the function f can be written as a pointwise limit of stepfunctions [compare Rudin (1976, chapter 6)], while stepfunctions are obviously simple functions. Refering to definition 2.1.7 we thus have:

Theorem 2.1.6. Let $\underline{x}_1, \ldots, x_k$ be r.v. on a probability space $(\Omega,\mathcal{F},P)$ and let $F(x_1, \ldots, x_k)$ be the joint distribution function of these r.v.'s. If for a Borel measurable function $f(x_1,\ldots,x_k)$ on R^k the Riemann-Stieltjes integral $\int_{R^k} f(x_1, \ldots, x_k)dF(x_1,\ldots,x_k)$ is defined then

$$E\ f(\underline{x}_1, \ldots, \underline{x}_k) = \int_{\Omega} f(x_1, \ldots, x_k)dP = \int_{R^k} f(x_1, \ldots, x_k)dF(x_1, \ldots,x_k).$$

The general integral with respect to a probability measure has all the properties of the classical Riemann-Stieltjes integral with respect to the distribution function F, and in the following we shall assume that the reader is familiar with its elementary properties. [Otherwise, see for example Chung (1974)]. Some of these properties are listed below since they are frequently used in this study. These properties are:

1) Let $\underline{x}$ and $\underline{y}$ be random variables defined on the probability space $\{\Omega,\mathcal{F},P\}$. Let Λ,Λ_n be sets in $\mathcal{F}$. Let α and β be real numbers. We have

(a) $\int_{\Lambda}(\alpha x + \beta y)dP=\alpha\int_{\Lambda} xdP + \beta\int_{\Lambda} ydP$

provided that the right side is meaningful.

(b) If the Λ_n's are disjoint, then

$$\int_{\bigcup_n \Lambda_n} xdP=\sum_n\int_{\Lambda_n} xdP$$

(c) If $x(\omega)\geq 0$ for every $\omega\in\Lambda$, then

$$\int_\Lambda x\,dP \geq 0$$

(d) If $x(\omega)\leq y(\omega)$ for every $\omega\in\Lambda$, then

$$\int_\Lambda x\,dP \leq \int_\Lambda y\,dP$$

(e) $$\left|\int_\Lambda x\,dP\right| \leq \int_\Lambda |x|\,dP$$

provided that the left side integral is defined.
These properties are not hard to verify from the definitions 2.1.7 and 2.1.8. See also Kolmogorov and Fomin (1961), chapter VII, for related properties of the Lebesgue integral.

2) If $\underline{x}_1, \underline{x}_2, \ldots, \underline{x}_n$ are independent random variables or -vectors and if $f_1(x),\ldots,f_n(x)$ are Borel measurable functions, then

$$E \prod_{j=1}^{n} f_j(\underline{x}_j) = \prod_{j=1}^{n} E f_j(\underline{x}_j) \qquad , n \leq \infty ,$$

provided that the right hand expectations are finite.

3) Chebishev's inequality. If ϕ is a Borel measurable function on R^1 such that $\phi(x)$ is positive and monotone increasing on $(0,\infty)$ and $\phi(x) = \phi(-x)$, then for every r.v. $\underline{x}$ and every $\delta > 0$ we have

$$P(|\underline{x}| > \delta) \leq \frac{E\,\phi(\underline{x})}{\phi(\delta)} .$$

4) Hölder's inequality. Let $\underline{x}$ and $\underline{y}$ be r.v.. Then

$$|E\,\underline{x}\,\underline{y}| \leq E\,|\underline{x}\,\underline{y}| \leq \{E\,|\underline{x}|^p\}^{\frac{1}{p}} . \{E\,|\underline{y}|^q\}^{\frac{1}{q}} .$$

for $p > 1$ and $\frac{1}{p} + \frac{1}{q} = 1$. (If $p = 2$ we have the well-known Cauchy - Schwarz-inequality.)

5) Liapounov's inequality. Let $\underline{x}$ be a r.v. Then for $1 \leq p \leq q \leq \infty$,

$$\{E|\underline{x}|^p\}^{\frac{1}{p}} \leq \{E|\underline{x}|^q\}^{\frac{1}{q}} .$$

(This follows straightforwardly from Hölder's inequality by putting $\underline{y}=1$ and replacing $\underline{x}$ with $|\underline{x}|^p$ and p with q/p.

Since assertion 2) is a standard result and the proof of Chebishev's inequality is nearly trivial, we shall only prove Hölder's inequality:

Without loss of generality we may assume $E|\underline{x}|^p > 0$ and $E|\underline{y}|^q > 0$ because otherwise the inequality is nearly trivial. From the convexity of the function $-\log x$ on $(0,\infty)$ it follows that for $a, b \in (0,\infty)$:

$$\frac{1}{p}\log a + \frac{1}{q}\log b \leq \log\left(\frac{1}{p}a + \frac{1}{q}b\right),$$

or equivalently $a^{\frac{1}{p}} b^{\frac{1}{q}} \leq \frac{1}{p}a + \frac{1}{q}b$. Substitute $a = \dfrac{|\underline{x}|^p}{E|\underline{x}|^p}$, $b = \dfrac{|\underline{y}|^p}{E|\underline{y}|^p}$.

Then by taking expectations we obtain the inequality involved.

Since the mean of a finite number of non-random variables in R^1 can be considered as a mathematical expectation, it follows from Hölder's inequality that for real numbers $x_1, \ldots, x_n, y_1, \ldots, y_n$:

$$\left|\frac{1}{n}\sum_{j=1}^n x_j \cdot y_j\right| \leq \left\{\frac{1}{n}\sum_{j=1}^n |x_j|^p\right\}^{\frac{1}{p}} \cdot \left\{\frac{1}{n}\sum_{j=1}^n |y_j|^q\right\}^{\frac{1}{q}}, \quad p > 1, \ \frac{1}{p} + \frac{1}{q} = 1 \tag{2.1.11}$$

and consequently, taking $y_j=1$,

$$\left|\sum_{j=1}^n x_j\right|^p \leq n^{p-1}\sum_{j=1}^n |x_j|^p, \quad p \geq 1. \tag{2.1.12}$$

This last inequality is a sharpening of the following trivial but useful inequality :

$$\left|\sum_{j=1}^n x_j\right|^p \leq n^p (\max |x_j|)^p \leq n^p \sum_{j=1}^n |x_j|^p, \quad p > 0. \tag{2.1.13}$$

Finally we shall also use the following result[see the proof of theorem 2.2.6].

Theorem 2.1.7. Let $\underline{x}$ be a r.v. on $\{\Omega, \mathcal{F}, P\}$ such that $E|\underline{x}| < \infty$. Let (Λ_n) be a sequence of sets in $\mathcal{F}$ such that $\lim_{n\to\infty} P(\Lambda_n) = 0$.

Then :

$$\lim_{n\to\infty} \int_{\Lambda_n} |x(\omega)| P(d\omega) = 0 .$$

Proof: Let $\Sigma_n = \{\omega \in \Omega : |x(\omega)| \geq n\}$ and $\Gamma_n = \{\omega \in \Omega : n \leq |x(\omega)| < n+1\}$.

From the fact that

$$\int_{\Lambda} |x(\omega)| P(d\omega) = \int_{\bigcup_{n=0}^{\infty}\Gamma_n} |x(\omega) dP(d\omega)| = \sum_{n=0}^{\infty} \int_{\Gamma_n} |x(\omega)| P(d\omega) < \infty$$

we conclude that

$$\int_{\Sigma_n} |x(\omega)|P(d\omega)=\sum_{k=n}^{\infty}\int_{\Gamma_k} |x(\omega)|P(d\omega)\to 0 \quad \text{for } n\to\infty$$

The theorem follows now from

$$\int_{\Lambda_n} |x(\omega)|P(d\omega) = \int_{\Lambda_n \cap \Sigma_k} |x(\omega)|P(d\omega) + \int_{\Lambda_n \cap \Sigma_k^c} |x(\omega)|P(d\omega)$$

$$\leq \int_{\Sigma_k} |x(\omega)|P(d\omega) + \int_{\Lambda_n \cap \Sigma_k^c} kP(d\omega) \leq \int_{\Sigma_k} |x(\omega)|P(d\omega) + kP(\Lambda_n) . \qquad \square$$

2.2. Convergence of random variables and distributions

2.2.1 Weak and strong convergence of random variables

In this section we shall deal with three important convergence concepts, namely convergence in probability, convergence almost surely and convergence in distribution. The first concept is well known:

Definition 2.2.1. Let $(\underline{x}_n)$ be a sequence of r.v.'s. We say that $\underline{x}_n$ converges in probability to a r.v. $\underline{x}$ if for every $\varepsilon > 0$

$$\lim_{n\to\infty} P(|\underline{x}_n - \underline{x}| \leq \varepsilon) = 1,$$

and we write: $\underline{x}_n \to \underline{x}$ in pr. or plim $\underline{x}_n = \underline{x}$.

However, a much stronger convergence concept is:

Definition 2.2.2. Let $(\underline{x}_n)$ be a sequence of r.v's on a common probability space $\{\Omega,\mathcal{F},P\}$. We say that $\underline{x}_n$ converges almost surely (a.s.) to a r.v. $\underline{x}$ (defined on $\{\Omega,\mathcal{F},P\}$) if there is a null set $N\in\mathcal{F}$ (that is a set in $\mathcal{F}$ satisfying $P(N)=0$) such that for every $\omega\in\Omega\setminus N$:

$$\lim_{n\to\infty} x_n(\omega)= x(\omega),$$

and we write : $\underline{x}_n\to \underline{x}$ a.s. or $\lim_{n\to\infty} \underline{x}_n = \underline{x}$ a.s.

A useful criterion for almost sure convergence of random variables is given by the following theorem.

<u>Theorem 2.2.1.</u> Let $\underline{x}$ and $\underline{x}_1, \underline{x}_2, \ldots$ be random variables on a common probability space $\{\Omega, \mathcal{F}, P\}$. Then $\underline{x}_n \to \underline{x}$ a.s. if and only if

$$\lim_{n\to\infty} P\left(\bigcap_{m=n}^{\infty} \{|\underline{x}_m - \underline{x}| \le \varepsilon\}\right) = 1 \quad \text{for every } \varepsilon > 0. \tag{2.2.1}$$

<u>Proof</u>: First we prove that $x(\omega) = \lim x_n(\omega)$ pointwise on $\Omega \setminus N$ implies (2.2.1). Let $\omega_0 \in \Omega \setminus N$. Then for every $\varepsilon > 0$ there is a number $n_0(\omega_0, \varepsilon)$ such that $|x_n(\omega_0) - x(\omega_0)| \le \varepsilon$ if $n \ge n_0(\omega_0, \varepsilon)$. Now consider the following set in $\mathcal{F}$:

$$A_n(\varepsilon) = \bigcap_{m=n}^{\infty} \{\omega \in \Omega : |x_n(\omega) - x(\omega)| \le \varepsilon\} .$$

Then $\omega_0 \in A(\varepsilon)_{n_0(\omega_0,\varepsilon)}$ and hence $\omega_0 \in \bigcup_{n=1}^{\infty} A_n(\varepsilon)$. Thus we have $\Omega \setminus N \subset \bigcup_{n=1}^{\infty} A_n(\varepsilon)$ and consequently $P(\Omega \setminus N) \le P(\bigcup_{n=1}^{\infty} A_n(\varepsilon))$. But $P(\Omega \setminus N) = P(\Omega) - P(N) = 1$ since N is a null set, hence $P(\bigcup_{n=1}^{\infty} A_n(\varepsilon)) = 1$.

Since $A_n(\varepsilon) \subset A_{n+1}(\varepsilon)$, we have $A_k(\varepsilon) = \bigcup_{n=1}^{k} A_n(\varepsilon)$ and thus

$$\lim_{k\to\infty} P(A_k(\varepsilon)) = \lim_{k\to\infty} P\left(\bigcup_{n=1}^{k} A_n(\varepsilon)\right) = P\left(\bigcup_{n=1}^{\infty} A_n(\varepsilon)\right) = 1 ,$$

which proves the first part of the theorem.

Next we prove that if $\lim_{n\to\infty} P(A_n(\varepsilon)) = 1$ for every $\varepsilon > 0$ then there is a null set N such that

$$x(\omega) = \lim x_n(\omega) \text{ pointwise on } \Omega \setminus N .$$

For every $\omega_0 \in \bigcup_{n=1}^{\infty} A_n(\varepsilon)$ there exists a index $n_0(\omega_0, \varepsilon)$ such that $\omega_0 \in A_{n_0(\omega_0,\varepsilon)}(\varepsilon)$ and thus $|x_n(\omega_0) - x(\omega_0)| \le \varepsilon$ if $n \ge n_0(\omega_0, \varepsilon)$ for every $\omega_0 \in \bigcup_{n=1}^{\infty} A_n(\varepsilon)$. Put $N_\varepsilon = \Omega \setminus \bigcup_{n=1}^{\infty} A_n(\varepsilon)$. Then $N_\varepsilon \in \mathcal{F}$ and

$P(N_\varepsilon) = P(\Omega) - P(\bigcup_{n=1}^{\infty} A_n(\varepsilon)) = 1 - 1 = 0$, hence N_ε is a null set.

However, we must find a null set that is independent of ε. Consider the sequence $\varepsilon_k = \frac{1}{k}$ with $k = 1, 2, \ldots$. We have just proved that for every k there is a null set $N_k^* = N_{\varepsilon_k}$ such that for every $\omega \in \Omega \setminus N_k^*$:

$$|x_n(\omega) - x(\omega)| \le \frac{1}{k} \text{ if } n \ge n_0(\omega_0, \tfrac{1}{k}) .$$

But then the same applies for every $\omega \in \Omega \setminus N$, where $N = \bigcup_{k=1}^{\infty} N_k^*$ is a countable union of null sets in $\mathcal{F}$ and hence itself a null set in $\mathcal{F}$. By putting $k = [\frac{1}{\varepsilon}]+1$ for an arbitrarily chosen $\varepsilon > 0$ we now see that for all $\omega \in \Omega \setminus N$:

$$|x_n(\omega) - x(\omega)| \leq \frac{1}{k} \leq \varepsilon \text{ if } n \geq n_0(\omega_0, \frac{1}{[\frac{1}{\varepsilon}]+1}) \quad ,$$

which proves the "only if" part of the theorem. □

From this theorem we see that

$\underline{x}_n \to \underline{x}$ a.s. implies $\underline{x}_n \to \underline{x}$ in pr.,

because $\bigcap_{m=n}^{\infty} \{|\underline{x}_m - \underline{x}| \leq \varepsilon\} \subset \{|\underline{x}_n - \underline{x}| \leq \varepsilon\}$ and consequently $P(\bigcap_{m=n}^{\infty} \{|\underline{x}_m - \underline{x}| \leq \varepsilon\}) \leq P(|\underline{x}_n - \underline{x}| \leq \varepsilon)$.

The following simple but important theorem provides another useful criterion for almost sure convergence.

Theorem 2.2.2. (Borel - Cantelli lemma)
Let $\underline{x}$ and $\underline{x}_1, \underline{x}_2, \ldots$ be r.v.'s. If for every $\varepsilon > 0$, $\sum_{n=1}^{\infty} P(|\underline{x}_n - \underline{x}| > \varepsilon) < \infty$ then $\underline{x}_n \to \underline{x}$ a.s..

Proof: Consider the set

$$A_n(\varepsilon) = \bigcap_{m=n}^{\infty} \{|\underline{x}_n - \underline{x}| \leq \varepsilon\} = \bigcap_{m=n}^{\infty} \{\omega \in \Omega: |\underline{x}_n(\omega) - \underline{x}(\omega)| \leq \varepsilon\},$$

where $\{\Omega, \mathcal{F}, P\}$ is the probability space involved. From theorem 2.2.1 it follows that it suffices to show $P(A_n(\varepsilon)) \to 1$ or equivalently, $P(A_n(\varepsilon))^c) \to 0$. But

$(A_n(\varepsilon))^c = \bigcup_{m=n}^{\infty} \{|\underline{x}_n - \underline{x}| > \varepsilon\}$, and hence $P((A_n(\varepsilon))^c) \leq \sum_{m=n}^{\infty} P(|\underline{x}_m - \underline{x}| > \varepsilon)$.

Since the latter sum is a tailsum of the convergent series $\sum_{n=1}^{\infty} P(|\underline{x}_n - \underline{x}| > \varepsilon)$ we must have $\sum_{m=n}^{\infty} P(|\underline{x}_n - \underline{x}| > \varepsilon) \to 0$ as $n \to \infty$, which proves the theorem. □

The a.s. convergence concept arises in a natural way from the strong laws of large numbers. Here we give two versions of these laws.

Theorem 2.2.3A. If $(\underline{x}_j)$ is a sequence of uncorrelated random variables and if

$E(\underline{x}_j - E\underline{x}_j)^2 = O(j^{\mu})$ for some $\mu < 1$,

then

$\frac{1}{n}\sum_{j=1}^{n}(\underline{x}_j - E\underline{x}_j) \to 0$ a.s.

Proof: This theorem is a further elaboration of the strong law of large numbers of Rademacher-Menchov [see Révész (1968), theorem 3.2.1 or Stout (1974), theorem

2.3.2], which states: Let $(\underline{y}_j)$ be a sequence of orthogonal random variables (orthogonality means that $E\underline{y}_{j_1}\underline{y}_{j_2}=0$ if $j_1\neq j_2$). If

$$\sum_{j=1}^{\infty}(\log j)E\underline{y}_j^2 < \infty$$

then

$$\sum_{j=1}^{\infty} \underline{y}_j \quad \text{converges a.s.}$$

(which means that $\underline{z}=\sum_{j=1}^{\infty} \underline{y}_j$ is a.s. a finite valued random variable).
Now let

$$\underline{y}_j=(\underline{x}_j - E\underline{x}_j)/j.$$

Then

$$\sum_{j=1}^{\infty}(\log j)E\underline{y}_j^2 = \sum_{j=1}^{\infty}(\log j)O(j^{\mu-2}) < \infty$$

for $\mu< 1$. (Because $\int_1^{\infty}(\log u)u^{\mu-2}du=\int_1^{\infty}(\log u)u^{\mu-1}d\log u = \int_0^{\infty} ve^{(\mu-1)v}dv < \infty$). Consequently

$$\sum_{j=1}^{\infty} \frac{\underline{x}_j - E\underline{x}_j}{j} \quad \text{converges a.s.}$$

From the Kronecker lemma [see Révész (1968), theorem 1.2.2 or Chung (1974), p.123] this implies that:

$$\frac{1}{n}\sum_{j=1}^{n}(\underline{x}_j - E\underline{x}_j) \to 0 \text{ a.s.} \qquad \square$$

<u>Remark</u>. The condition $E(\underline{x}_j - E\underline{x}_j)^2 = O(j^{\mu})$ for some $\mu< 1$ holds if

$$\sup_n \frac{1}{n}\sum_{j=1}^{n}E|\underline{x}_j - E\underline{x}_j|^{2+\delta} < \infty \quad \text{for some } \delta>0,$$

because then $E|\underline{x}_j - E\underline{x}_j|^{2+\delta}=O(j)$, so that by Liapounov's inequality

$$E(\underline{x}_j - E\underline{x}_j)^2 \leq \{E|\underline{x}_j - E\underline{x}_j|^{2+\delta}\}^{\frac{2}{2+\delta}} = O(j^{\frac{2}{2+\delta}}).$$

If the $\underline{x}_j$'s are independent and equally distributed, the condition on the second moments is not needed:

<u>Theorem 2.2.3B:</u> (<u>Strong law of large numbers of Kolmogorov</u>)

If $(\underline{x}_j)$ is a sequence of independent and equally distributed random variables with $E|\underline{x}_1| < \infty$, then $\frac{1}{n}\sum_{j=1}^{n} \underline{x}_j \to E\,\underline{x}_1$ a.s.

Since in this study this theorem is only used in the discussion of the work of Jennrich (1969) in the next chapter, and since it does not play a further role in our own contributions to nonlinear estimation, we refer for the proof to Chung (1974), theorem 5.4.2.

We have already mentioned that almost sure convergence implies convergence in probability. There is also a converse connection, given by the following theorem.

Theorem 2.2.4. Let $\underline{x}$ and $\underline{x}_1, \underline{x}_2, \ldots$ be r.v's. Then $\underline{x}_n \to \underline{x}$ in pr. if and only if every subsequence (n_k) of the sequence (n) contains a further subsequence (n_{k_j}) such that $\underline{x}_{n_{k_j}} \to \underline{x}$ a.s.

Proof: Suppose that every subsequence (n_k) contains a further subsequence (n_{k_j}) such that $\underline{x}_{n_{k_j}} \to \underline{x}$ a.s., but that not $\underline{x}_n \to \underline{x}$ in pr. Then there exist numbers $\varepsilon > 0$, $\delta > 0$, and a subsequence (n_k) such that

$$P(|\underline{x}_{n_k} - \underline{x}| \leq \varepsilon) \leq 1 - \delta,$$

hence for every further subsequence (n_{k_j}) we have the same, which contradicts our assumption. Thus the "if" part is now proved. Next we suppose that $\underline{x}_n \to \underline{x}$ in pr. Then for every positive integer k

$$\lim_{n\to\infty} p(|\underline{x}_n - \underline{x}| > \frac{1}{2^k}) = 0.$$

For each k we can find a n_k such that $P(|\underline{x}_{n_k} - \underline{x}| > \frac{1}{2^k}) \leq \frac{1}{2^k}$, hence

$$\sum_{k=0}^{\infty} P(|\underline{x}_{n_k} - \underline{x}| > \frac{1}{2^k}) \leq \sum_{k=1}^{\infty} \frac{1}{2^k} < \infty$$

and consequently

$$\sum_{k=0}^{\infty} P(|\underline{x}_{n_k} - \underline{x}| > \varepsilon) < \infty \text{ for every } \varepsilon > 0.$$

By the Borel-Cantelli lemma it follows now that $\underline{x}_{n_k} \to \underline{x}$ a s., which proves the "only if" part. □

Theorem 2.2.5. Let $\underline{x}$ and $\underline{x}_1, \underline{x}_2, \ldots$ be r.v's such that

a) $\underline{x}_n \to \underline{x}$ a.s.

or

b) $\underline{x}_n \to \underline{x}$ in pr.,

respectively. Let f(x) be a Borel measurable function on R^1. If f is continuous on a Borel set B such that $P(\underline{x} \in B) = 1$, then

a) $f(\underline{x}_n) \to f(\underline{x})$ a.s.

or

b) $f(\underline{x}_n) \to f(\underline{x})$ in pr.,

respectively .

Proof: a) Let $\{\Omega,\mathcal{F},P\}$ be the probability space involved. There is a null set N_1 such that for every $\varepsilon > 0$ and every $\omega \in \Omega \setminus N_1$:

$$|x_n(\omega) - x(\omega)| \leq \varepsilon \quad \text{if } n \geq n_0(\omega,\varepsilon) .$$

Let $N_2 = \Omega \setminus \{ \omega\in\Omega \ : x(\omega) \in B \}$. Then N_2 is a null set in $\mathcal{F}$ and $x(\omega)$ is for every $\omega \in \Omega \setminus N_2$ a continuity point of f. Thus for every $\varepsilon > 0$ and every $\omega \in \Omega \setminus N_2$ there is a number $\delta(\varepsilon,\omega)$ such that:

$$|f(x_n(\omega)) - f(x(\omega))| \leq \varepsilon \quad \text{if } |x_n(\omega) - x(\omega)| \leq \delta(\varepsilon,\omega) .$$

Hence for every $\varepsilon > 0$ and every $\omega \in \Omega\setminus(N_1 \cup N_2)$ we have:

$$|f(x_n(\omega)) - f(x(\omega))| \leq \varepsilon \text{ if } n \geq n_0(\omega, \delta(\varepsilon,\omega)).$$

Since $N_1 \cup N_2$ is a null set in $\mathcal{F}$, this proves part a) of the theorem.

b) For an arbitrary subsequence (n_k) we have a further subsequence (n_{k_j}) such that $\underline{x}_{n_k} \to \underline{x}$ a.s.(see theorem 2.2.4) and consequently by part a) of the theorem: $f(\underline{x}_{n_{k_j}}) \to f(\underline{x})$ a.s. By theorem 2.2.4 this implies $f(\underline{x}_n) \to f(\underline{x})$ in pr. □

Remark. Until so far in this section we only dealt with random variables in R^1. But generalisation of the definitions and the theorems in this section from random variables to finite dimensional random vectors is straightforward simply by changing random variable to random vector.

2.2.2 Convergence of mathematical expectations

If $\underline{x}$ and $\underline{x}_1, \underline{x}_2, \ldots$ are random variables or vectors such that for some $p > 0$, $E|\underline{x}_n - \underline{x}|^p \to 0$ as $n \to \infty$, then it follows from Chebishev's inequality that $\underline{x}_n \to \underline{x}$ in pr. The converse is not always true. A partial converse is given by the following theorem.

Theorem 2.2.6. If $\underline{x}_n \to \underline{x}$ in pr. and if there is a r.v. $\underline{y}$ satisfying $|\underline{x}_n| \leq \underline{y}$ a.s. for $n = 1,2, \ldots$ and $E\,\underline{y}^p < \infty$ for some $p > 0$, then: $E\,|\underline{x}_n - \underline{x}|^p \to 0$.

Proof. If $P(|\underline{x}| > \underline{y}) > 0$ then $\underline{x}_n \to \underline{x}$ in pr. is not possible. Hence $|\underline{x}| \leq \underline{y}$ a.s. Since now $|\underline{x}_n - \underline{x}| \leq 2\,\underline{y}$ there is no loss of generality in assuming $\underline{x} = 0$ a.s..

We then have:

$$\int|x_n(\omega)|^p P(d\omega)=\int_{\{|x_n(\omega)|>\varepsilon\}}|x_n(\omega)|^p P(d\omega)+\int_{\{|x_n(\omega)|\leq\varepsilon\}}|x_n(\omega)|^p P(d\omega)\leq\varepsilon^p+\int_{\{|x_n(\omega)|>\varepsilon\}}y(\omega)^p P(d\omega).$$

The theorem follows now from theorem 2.1.7. □

Putting p=1 in theorem 2.2.6 we have:

Theorem 2.2.7. (Bounded convergence theorem)

If $\underline{x}_n \to \underline{x}$ in pr. and if $|\underline{x}_n| \leq \underline{y}$ a.s., where $E\,\underline{y} < \infty$, then $E\,\underline{x}_n \to E\,\underline{x}$

We use this theorem for proving first Fatou's lemma which in its turn will be used for proving the monotone convergence theorem.

Theorem 2.2.8. (Fatou's lemma)

If $\underline{x}_n \geq 0$ a.s., then $E \liminf \underline{x}_n \leq \liminf E\,\underline{x}_n$.

Proof: Put $\underline{x} = \liminf \underline{x}_n$ and let $\phi(x)$ be any simple function satisfying $0 \leq \phi(x) \leq x$. Put $\underline{y}_n = \min(\phi(\underline{x}), \underline{x}_n)$. Then $\underline{y}_n \to \phi(\underline{x})$ in pr. because:

$P(|\min(\phi(\underline{x}), \underline{x}_n) - \phi(\underline{x})| \geq \varepsilon) = P(\underline{x}_n \leqslant \phi(\underline{x}) - \varepsilon) \leq P(\underline{x}_n \leqslant \underline{x} - \varepsilon) \to 0.$

Moreover, since $\phi(x)$ is a simple function we must have $E\phi(\underline{x}) < \infty$. From the bounded convergence theorem and from $\underline{y}_n \leq \underline{x}_n$ a.s. it follows now:

$E\,\phi(\underline{x}) = \lim E\,\underline{y}_n = \liminf E\,\underline{y}_n \leq \liminf E\,\underline{x}_n.$

Taking the supremum over all such simple functions ϕ it follows now from definition 2.1.7:

$E\,\underline{x} \leq \liminf E\,\underline{x}_n,$

which was to be proved. □

Theorem 2.2.9. (Monotone convergence theorem.)

Let $(\underline{x}_n)$ be a non-decreasing sequence of r.v.'s. Then $E \lim \underline{x}_n = \lim E\,\underline{x}_n \leq \infty$.

Proof: Since our sequence $(\underline{x}_n)$ is non-decreasing, we have:

$\lim \underline{x}_n = \liminf \underline{x}_n$, $\lim E\,\underline{x}_n = \liminf E\,\underline{x}_n$,

so that by Fatou's lemma:

$E \lim \underline{x}_n \leq \lim E\,\underline{x}_n$.

But for any n we have $\underline{x}_n \leq \lim \underline{x}_n$ a.s. because $\underline{x}_n$ is non-decreasing, hence $E\,\underline{x}_n \leq E \lim \underline{x}_n$ and consequently:

$\lim E\,\underline{x}_n \leq E \lim \underline{x}_n$,

which proves the theorem. □

2.2.3 Convergence of distributions

If $\underline{x}$ and $\underline{x}_1, \underline{x}_2, \ldots$ are r.v's with distribution functions F, F_1, $F_2, \ldots$, respectively, then one would like to say that $\underline{x}_n$ convergences in distribution to $\underline{x}$ if for every $t \in R^1$, $F_n(t) \to F(t)$.
However, if $\underline{x}$ and F are given and if we define $\underline{x}_n = \underline{x} + \frac{1}{n}$, then:

$$F_n(t) = P(\underline{x}_n \leq t) = P(\underline{x} \leq t - \frac{1}{n}) = F(t - \frac{1}{n}),$$

so that for every discontinuity point t_0 of F we have :

$$\lim F_n(t_0) = \lim F(t_0 - \frac{1}{n}) = F(t_0 -) < F(t_0),$$

while intuitively we should expect that in this case we also have convergence in distribution. Moreover, if $\underline{x}_n = \underline{x} + n$ we have $F_n(t) = P(\underline{x}_n \leq t) =$ $F(t-n) \to F(-\infty) = 0$ for every t. Thus not every sequence of distribution functions converges to another distribution function. In the latter case we say that the convergence is improper.

Definition 2.2.3. A sequence $(F_n(t))$ of distribution functions converges properly if $F_n(t) \to F(t)$ pointwise for all continuity points of F, where F is a distribution function. We then write: $F_n \to F$ properly.

The exclusion of discontinuity points avoids the complication that otherwise the function $F(t) = \lim F_n(t)$ may not be right continuous. In view of the above example we now define:

Definition 2.2.4. A sequence $(\underline{x}_n)$ of random variables (or random vectors) converges in distribution to a random variable (or random vector) $\underline{x}$, if their underlying distribution functions F_n, F, respectively, satisfy $F_n \to F$ properly. We then write $\underline{x}_n \to \underline{x}$ in distr..

Remark: If this "limit" distribution F is the distribution function of (for example) the normal distribution $N(\mu,\sigma^2)$, we shall also write: $\underline{x}_n \to N(\mu,\sigma^2)$ in distr.

There is a close connection between proper convergence of distribution functions and convergence of mathematical expectations, as is shown by the following theorem. This theorem is very fundamental since it allows a variety of applications.

Theorem 2.2.10. Let F and F_n, n = 1,2, ... be distribution functions on R^k. Then $F_n \to F$ properly if and only if for every uniformly bounded continuous real function ϕ on R^k:

$$\int_{R^k} \phi(t)dF_n(t) \to \int_{R^k} \phi(t)dF(t) .$$

Proof: Since the proof for the general case k > 1 is a straightforward extension of that for k = 1, we assume k = 1.
Suppose $F_n \to F$ properly. We can always find for given $\varepsilon>0$ continuity points a and b of F such that $F(b)-F(a)>1-\varepsilon$. Let ϕ be any uniformly bounded continuous real function on R with uniform bound 1 (which is no restriction). By the uniform continuity of ϕ on [a,b] we can find continuity points $t_2, t_3, \ldots, t_{m-1}$ of F satisfying

$$a=t_1<t_2< \ldots < t_{m-1}<b=t_m \text{ and } \sup_{t\in(t_i,t_{i+1}]} \phi(t) - \inf_{t\in(t_i,t_{i+1}]} \phi(t) \le \varepsilon \text{ for } i=1,2,\ldots,m-1.$$

Now define

$$\psi(t)=\begin{cases} \inf_{t\in(t_i,t_{i+1}]} \phi(t) & \text{for } t\in(t_i,t_{i+1}],\ i=1,2,\ldots,m-1. \\ 0 & \text{elsewhere}. \end{cases}$$

Then

$$0\le \phi(t)-\psi(t) \le \varepsilon \qquad \text{for } t\in (a,b],$$

$$0\le \phi(t)-\psi(t) \le 1 \qquad \text{for } t\notin (a,b],$$

hence:

$$\left| \int \psi(t)dF_n(t)- \int \phi(t)dF_n(t)\right| \le \int_{\{t\in(a,b]\}} \varepsilon dF_n(t)+ \int_{\{t\notin(a,b]\}} dF_n(t)$$

$$= \varepsilon(F_n(b)-F_n(a))+1-F_n(b)+F_n(a)$$

$$\to \varepsilon(F(b)-F(a))+1-F(b)+F(a) \le 2\varepsilon.$$

Moreover,

$$\int \psi(t)dF_n(t)=\sum_{i=1}^{m-1}\{\inf_{t\in(t_i,t_{i+1}]}\phi(t)\}(F_n(t_{i+1})-F_n(t_i))$$

$$\to\sum_{i=1}^{m-1}\{\inf_{t\in(t_i,t_{i+1}]}\phi(t)\}(F(t_{i+1})-F(t_i))=\int\psi(t)dF(t)$$

and $|\int \phi(t)\,dF(t)-\int \psi(t)dF(t)| \leq 2\varepsilon$. So we have:

$$|\int \phi(t)\,dF_n(t)-\int \phi(t)dF(t)| \leq 4\varepsilon + |\int \psi(t)dF_n(t)-\int \psi(t)dF(t)| \leq 5\,\varepsilon$$

for sufficiently large n, which proves the "only if" part of the theorem.

Now let u be a continuity point of F and define

$$\phi(t) = \begin{cases} 1 \text{ if } t \leq u \\ 0 \text{ if } t > u, \end{cases}$$

$$\phi_{1,m}(t) = \begin{cases} 1 \text{ if } t \leq u - \frac{1}{m} \\ - m.t + m.u \text{ if } t \in (u - \frac{1}{m}, u] \\ 0 \text{ if } t > u, \end{cases}$$

$$\phi_{2,m}(t) = \begin{cases} 1 \text{ if } t \leq u \\ - m.t + 1 + m.u \quad \text{if } t \in (u, u + \frac{1}{m}] \\ 0 \quad \text{if } t > u + \frac{1}{m}. \end{cases}$$

Then $\phi_{1,m}$ and $\phi_{2,m}$ are uniformly bounded continuous functions on R^1 satisfying $\phi_{1,m}(t) \leq \phi(t) \leq \phi_{2,m}(t)$, hence for m = 1,2,... and $n \to \infty$:

$$\begin{array}{ccccc} \int \phi_{1,m}(t)dF_n(t) \leq & F_n(u)= & \int \phi(t)dF_n(t) & \leq & \int \phi_{2,m}(t)dF_n(t) \\ \downarrow & & & & \downarrow \\ \int \phi_{1,m}(t)dF(t) \leq & F(u)= & \int \phi(t)dF(t) & \leq & \int \phi_{2,m}(t)dF(t). \end{array}$$

Moreover

$$0 \leq \int(\phi_{2,m}(t)-\phi_{1,m}(t))dF(t) \leq \int_{\{t\in(u-\frac{1}{m},\,u+\frac{1}{m}]\}} dF(t) \leq F(u+\tfrac{1}{m})-F(u-\tfrac{1}{m}).$$

Since u is a continuity point of F, $F(u+\frac{1}{m}) - F(u-\frac{1}{m})$ can be made arbitrarily small by increasing m; hence $F_n(u) \to F(u)$, which proves the "if" part. □

A direct consequence of this theorem is that

Theorem 2.2.11. $\underline{x}_n \to \underline{x}$ in pr. implies $\underline{x}_n \to \underline{x}$ in distr.,

because by theorem 2.2.5 it follows that for any continuous function ϕ we have that $\underline{x}_n \to \underline{x}$ in pr. implies $\phi(\underline{x}_n) \to \phi(\underline{x})$ in pr., while by theorem 2.2.7, $\phi(\underline{x}_n) \to \phi(\underline{x})$ in pr. implies $E\phi(\underline{x}_n) \to E\phi(\underline{x})$ if ϕ is a uniformly bounded continuous function.

The converse of this theorem is not generally true, but it is if $\underline{x}$ is constant a.s., that is: $P(\underline{x}=c)=1$ for some constant c. In that case the proper limit F involved is:

$$F(t) = \begin{cases} 1 & \text{if } t \geq c \\ 0 & \text{if } t < c \end{cases} .$$

The proof of this proposition is very simple:

$$P(|\underline{x}_n - c| \leq \varepsilon) = P(c-\varepsilon \leq \underline{x}_n \leq c+\varepsilon) = F_n(c+\varepsilon) - F_n((c-\varepsilon)-)$$
$$\to F(c+\varepsilon) - F(c-\varepsilon) = 1 \quad \text{for every } \varepsilon > 0 ,$$

since $c+\varepsilon$ and $c-\varepsilon$ are continuity points of F. Thus:

Theorem 2.2.12. Convergence in distribution to a constant implies convergence in probability to that constant.

Let $\underline{x}_n$ and $\underline{x}$ be random vectors in R^k such that $\underline{x}_n \to \underline{x}$ in distr. and let f be any continuous real function on R^k. For any uniformly bounded continuous real function ϕ on R^1 it follows that $\phi(f)$ is a uniformly bounded continuous real function on R^k, so that by theorem 2.2.10, $E\phi(f(\underline{x}_n)) \to E\phi(f(\underline{x}))$ and consequently $f(\underline{x}_n) \to f(\underline{x})$ in distr. Thus we have:

Theorem 2.2.13. Let $\underline{x}_n$ and $\underline{x}$ be random vectors in R^k and let f be a continuous real function on R^k. Then $\underline{x}_n \to \underline{x}$ in distr. implies $f(\underline{x}_n) \to f(\underline{x})$ in distr..

A more general result is given by the following theorem.

Theorem 2.2.14. Let $\underline{x}_n$ and $\underline{x}$ be random vectors in R^k, $\underline{y}_n$ a random vector in R^m and c a non-random vector in R^m.
If $\underline{x}_n \to \underline{x}$ in distr. and $\underline{y}_n \to c$ in distr. then for any continuous real function f on $R^k \times C$,where C is some subset of R^m with interior point c, $f(\underline{x}_n,\underline{y}_n) \to f(\underline{x},c)$ in distr.

Proof: Again we prove the theorem for the case k=m=1 since the proof of the general case is similar.
It suffices to prove that for any uniformly bounded continuous real function ϕ on R^2 we have

$$E\phi(\underline{x}_n,\underline{y}_n) \to E\phi(\underline{x},c),$$

because then

$$E\psi(f(\underline{x}_n,\underline{y}_n)) \to E\psi(f(\underline{x},c))$$

for any uniformly bounded continuous real function ψ on R^1, which by theorem 2.2.10 implies $f(\underline{x}_n,\underline{y}_n) \to f(\underline{x},c)$ in distr.
Let M be the uniform bound of ϕ and let F_n and F be the distribution functions of $\underline{x}_n$ and $\underline{x}$, respectively. For every ε we can choose continuity points a and b of F such that

$$P(\underline{x}\in(a,b]) = F(b) - F(a) > 1 - \frac{\varepsilon}{2M}$$

Moreover, for any $\delta > 0$ we have

$$|E\phi(\underline{x}_n,\underline{y}_n) - E\phi(\underline{x}_n,c)| \leq \left| \int_{\{\underline{x}_n\in(a,b]\}} |\phi(\underline{x}_n,\underline{y}_n) - \phi(\underline{x}_n,c)|dP \right. +$$

$$+ \int_{\{\underline{x}_n\notin(a,b]\}} |\phi(\underline{x}_n,\underline{y}_n) - \phi(\underline{x}_n,c)|dP \leq$$

$$\leq \int_{\{\underline{x}_n\in(a,b]\}\cap\{|\underline{y}_n - c|\leq \delta\}} |\phi(\underline{x}_n,\underline{y}_n) - \phi(\underline{x}_n,c)|dP + 2M\, P(\{\underline{x}_n\in(a,b]\}\cap\{|\underline{y}_n - c|>\delta\}) + 2M\, P(\underline{x}_n\notin(a,b]) \leq$$

$$\leq \int_{\{\underline{x}_n\in(a,b]\}\cap\{|\underline{y}_n-c|\leq\delta\}} |\phi(\underline{x}_n,\underline{y}_n) - \phi(\underline{x}_n,c)|dP + 2M\, P\{|\underline{y}_n-c|>\delta\} + 2M(1-F_n(b)+F_n(a))$$

Since by theorem 2.2.12, $\underline{y}_n \to c$ in pr., we have $P\{|\underline{y}_n - c| > \delta\} \to 0$ for any $\delta>0$, while $\lim 2M(1-F_n(b)+F_n(a)) = 2M(1-F(b)+F(a)) < \varepsilon$. Furthermore, since $\phi(t_1,t_2)$ is uniformly continuous on the bounded set $\{(t_1,t_2)\in R^2 : a < t_1 \leq b, |t_2 - c| \leq \delta\}$, provided that δ is so small that this set is contained in C, we can choose δ such that the last written integral above is smaller than ε. So we conclude:

$$E\phi(\underline{x}_n,\underline{y}_n) - E\phi(\underline{x}_n,c) \to 0.$$

Since obviously $E\phi(\underline{x}_n,c) \to E\phi(\underline{x},c)$ because $\underline{x}_n \to \underline{x}$ in distr., the theorem follows. □

2.2.4 <u>Convergence of distributions and mathematical expectations</u>

The condition in theorem 2.2.10 that the function ϕ is uniformly bounded is a serious limitation for econometric applications. So we shall try to get rid of it. But before doing this we introduce some notation:

<u>Definition 2.2.5</u>: Let ϕ be any function on a Euclidean space. By $\overline{\phi(x)}^{c}$ we mean the function

$$\overline{\phi(x)}^{o} = \sup_{\{|y|\leq|x|\}} |\phi(y)|, \tag{2.2.1}$$

which obviously is a function of $|x|$. This notation plays a role in the following extension of the "only if" part of theorem 2.2.10.

<u>Theorem 2.2.15</u>. Let (F_n) be a sequence of distribution functions on R^k satisfying $F_n \to F$ properly. Let $\phi(x)$ be a continuous real function on R^k such that

$$\sup_n \int \{\overline{\phi(x)}^{o}\}^{1+\delta} dF_n(x) < \infty \text{ for some } \delta > 0. \tag{2.2.2}$$

Then

$$\int \phi(x)dF_n(x) \to \int \phi(x)dF(x).$$

For proving this theorem we need the following lemma.

<u>Lemma 2.2.1</u>. Let ϕ be a continuous real function on R^k and let for $a \geq 0$:

$$\psi(a) = \sup_{|x|\leq a} |\phi(x)|.$$

Then $\psi(a)$ is continuous on $[0,\infty)$.

<u>Proof</u>: The function ϕ is uniformly continuous on the set $S=\{x\in R^k : |x| \leq a+1\}$, $a > 0$. Thus for every $\varepsilon>0$ there is a $\delta>0$ such that for every $x_1\in S$, $x_2\in S$ satisfying $|x_1 - x_2| < \delta$ we have $|\phi(x_1) - \phi(x_2)| < \varepsilon$. Now let $x\in S$ and let θ be a number in $(0,1)$.

Then $\theta x \in S$ and

$$|\phi(x)-\phi(\theta x)|<\varepsilon \quad \text{if } |x-\theta x|=(1-\theta)|x|\leq(1-\theta)(a+1)<\delta .$$

Thus choose θ such that

$$1-\frac{\delta}{a+1}<\theta<1,$$

which is possible for θ sufficiently close to 1.
Next let $0<\delta_1<1$. We have

$$\psi(a+\delta_1) = \sup_{|x|\leq a+\delta_1}|\phi(x)|\leq \sup_{|x|\leq a+\delta_1}|\phi(x)-\phi(\theta x)|+\sup_{|x|\leq a+\delta_1}|\phi(\theta x)|<$$

$$<\varepsilon+\sup_{|x|\leq\theta(a+\delta_1)}|\phi(x)|\leq\varepsilon+\psi(a)$$

if $\theta(a+\delta_1)\leq a$. Thus if we choose δ_1 so small that

$$1-\frac{\delta}{a+1}<\theta<\frac{a}{a+\delta_1},$$

then

$$0\leq\psi(b)-\psi(a)\leq\varepsilon \quad \text{for } a<b<a+\delta_1.$$

This proves that ψ is left continuous on $(0,\infty)$. By letting $a\downarrow 0$ we see that ψ is also left continuous on $[0,\infty)$. By a similar argument it can be shown that ψ is right continuous on $(0,\infty)$. □

<u>Proof of theorem 2.2.15</u>. Since $|\phi(x)|\leq\overline{\phi(x)}^{o}$ and $\overline{\phi(x)}^{o}$ is a non-decreasing function of $|x|$ it follows that $\lim_{|x|\to\infty}\overline{\phi(x)}^{o}<\infty$ implies that $\phi(x)$ is uniformly bounded. In that case the theorem follows from theorem 2.2.10. Thus without loss of generality we may assume

$$\overline{\phi(x)}^{o}\to\infty \quad \text{as } |x|\to\infty. \tag{2.2.3}$$

Now define for $a>0$:

$$\phi_a(x)=\begin{cases}\phi(x) & \text{if } |x|\leq a.\\ \phi(\frac{a}{|x|}x) & \text{if } |x|>a\end{cases} \tag{2.2.4}$$

Then $\phi_a(x)$ is a uniformly bounded continuous real function on R^k, hence by theorem 2.2.10:

$$\int\phi_a(x)dF_n(x)\to\int\phi_a(x)dF(x). \tag{2.2.5}$$

Moreover,

$$\left|\int\phi(x)dF_n(x) - \int\phi_a(x)dF_n(x)\right| = \left|\int_{\{|x|>a\}}\phi(x)dF_n(x) - \int_{\{|x|>a\}}\phi_a(x)dF_n(x)\right|$$

$$\leq 2\int_{\{|x|>a\}}\overline{\phi(x)}^{\circ}dF_n(x) = 2\int_{\{|x|>a\}}\frac{\{\overline{\phi(x)}^{\circ}\}^{\delta+1}}{\{\overline{\phi(x)}^{\circ}\}^{\delta}}dF_n(x)$$

$$\leq \frac{2}{\inf\limits_{\{|x|>a\}}\{\overline{\phi(x)}^{\circ}\}^{\delta}}\sup_n\int\{\overline{\phi(x)}^{\circ}\}^{1+\delta}dF_n(x) \to 0 \quad \text{as } a\to\infty, \tag{2.2.6}$$

because of (2.2.2) and (2.2.3). Similarly we have

$$\left|\int\phi(x)dF(x) - \int\phi_a(x)dF(x)\right| \leq 2\int_{\{|x|>a\}}\overline{\phi(x)}^{\circ}dF(x). \tag{2.2.7}$$

If $\int\overline{\phi(x)}^{\circ}dF(x)$ converges, then (2.2.7) tends to zero as $a\to\infty$, hence the theorem follows from (2.2.5), (2.2.6) and (2.2.7). For showing

$$\int\overline{\phi(x)}^{\circ}dF(x) < \infty \tag{2.2.8}$$

we observe that by lemma 2.2.1, $\psi(|x|)=\overline{\phi(x)}^{\circ}$ is a continuous function of $|x|$. Now let

$$\psi_a(|x|)=\begin{cases}\psi(|x|) & \text{if } |x|\leq a\\ \psi(a) & \text{if } |x|>a.\end{cases} \tag{2.2.9}$$

Then $\psi_a(|x|)$ is a uniformly bounded continuous function on R^k, and for any $|x|$ a non decreasing function of $a>0$. Since obviously $\psi_a(|x|)\leq\psi(|x|)$, and $\lim\limits_{a\to\infty}\psi_a(|x|)=\psi(|x|)$, we have by the monotone convergence theorem and theorem 2.2.10:

$$\int\psi(|x|)dF(x) = \lim_{a\to\infty}\int\psi_a(|x|)dF(x) = \lim_{a\to\infty}\{\lim_{n\to\infty}\int\psi_a(|x|)dF_n(x)\}$$

$$\leq \sup_n\int\psi(|x|)dF_n(x) = \sup_n\int\overline{\psi(x)}^{\circ}dF_n(x) < \infty. \tag{2.2.10}$$

Thus (2.2.8) is proved by now and so is the theorem. □

Along similar lines we can prove the following version of the well-known weak law of large numbers:

Theorem 2.2.16. Let $\underline{x}_1, \underline{x}_2, \ldots$ be a sequence of independent random vectors in R^k, and let $(F_j(x))$ be the sequence of corresponding distribution functions. Let $\phi(x)$ be a continuous function on R^k. If

$$\frac{1}{n}\sum_{j=1}^{n} F_j(x) \to G(x) \text{ properly}$$

and

$$\sup_n \frac{1}{n}\sum_{j=1}^{n} E\left\{\overline{\phi(\underline{x}_j)}^{o}\right\}^{1+\delta} < \infty \text{ for some } \delta > 0,$$

then

$$\text{plim}\ \frac{1}{n}\sum_{j=1}^{n} \phi(\underline{x}_j) = \int \phi(x)dG(x).$$

Proof: If $\overline{\phi(x)}^{o}$ remains bounded for $|x| \to \infty$, then $\phi(x)$ is uniformly bounded. In that case the theorem follows from the usual weak law of large numbers and theorem 2.2.10. Thus we assume now that $\overline{\phi(x)}^{o} \to \infty$ as $|x| \to \infty$.

Consider the function $\phi_a(x)$ defined in (2.2.4). Then obviously by the independence of the $\underline{x}_j$'s, the boundedness of $\phi_a(x)$ and Chebishev's inequality:

$$\text{plim}_{n\to\infty} \frac{1}{n}\sum_{j=1}^{n}\{\phi_a(\underline{x}_j) - E\ \phi_a(\underline{x}_j)\} = 0, \tag{2.2.11}$$

while from theorem 2.2.10 it follows

$$\lim_{n\to\infty} E\frac{1}{n}\sum_{j=1}^{n}\phi_a(\underline{x}_j) = \int \phi_a(x)dG(x). \tag{2.2.12}$$

Hence

$$\text{plim}_{n\to\infty} \frac{1}{n}\sum_{j=1}^{n}\phi_a(\underline{x}_j) = \int \phi_a(x)dG(x). \tag{2.2.13}$$

Moreover, since $\phi_a(x)$ is uniformly bounded, it follows from theorem 2.2.6 that (2.2.13) implies

$$E\left|\frac{1}{n}\sum_{j=1}^{n}\phi_a(\underline{x}_j) - \int \phi_a(x)dG(x)\right| \to 0 \quad \text{as } n \to \infty. \tag{2.2.14}$$

Furthermore, similar to (2.2.6) it follows that

$$\limsup_n E\left|\frac{1}{n}\sum_{j=1}^{n}\phi(\underline{x}_j) - \frac{1}{n}\sum_{j=1}^{n}\phi_a(\underline{x}_j)\right| \to 0 \quad \text{as} \quad a \to \infty \tag{2.2.15}$$

and similar to (2.2.7) that

$$\left|\int \phi(x)dG(x) - \int \phi_a(x)dG(x)\right| \to 0 \quad \text{as} \quad a \to \infty. \tag{2.2.16}$$

Combining (2.2.14), (2.2.15) and (2.2.16) we see that

$$\lim E\left|\frac{1}{n}\sum_{j=1}^{n}\phi(\underline{x}_j) - \int\phi(x)dG(x)\right| \to 0. \tag{2.2.17}$$

The theorem follows now from (2.2.17) and Chebishev's inequality. □

Remark. The difference of this theorem with the classical weak law of large numbers is that the finiteness of second moments is not necessary.

If we combine the theorems 2.2.3A and 2.2.15 we obtain the following strong version of theorem 2.2.16.

Theorem 2.2.17. Let the conditions of theorem 2.2.16 be satisfied, and assume in addition that

$$E\{\overline{\phi(\underline{x}_j)}^{\circ}\}^2 = O(j^{\mu}) \text{ with } \mu < 1 \;^{*)} \tag{2.2.18}$$

Then $\frac{1}{n}\sum_{j=1}^{n}\phi(\underline{x}_j) \to \int\phi(x)dG(x)$ a.s.

2.3 Uniform convergence of random functions

2.3.1 Random functions. Uniform strong and weak convergence

A random function is a function which is a r.v. for each value of its argument. Usually random functions occur as a function of both random variables and parameters, for example the sum of squares of a regression model. Their definition is similar to that of random variables:

Definition 2.3.1: Let $\{\Omega, \mathfrak{F}, P\}$ be a probability space and let Ⓗ be a subset of R^k. The real function $\underline{f}(\theta)=f(\theta,\omega)$ on Ⓗ x Ω is called a (real) random function on Ⓗ if for every $t\in R^1$ and every $\theta\in$ Ⓗ, $\{\omega\in\Omega: f(\theta,\omega) \leq t\}\in\mathfrak{F}$.

*) Note that $E|\phi(\underline{x}_j) - E\phi(\underline{x}_j)|^2 \leq E\{\overline{\phi(x_j)}^{\circ}\}^2$. The condition $E|\phi(\underline{x}_j) - E\phi(\underline{x}_j)|^2 = O(j^{\mu})$ with $\mu<1$ would be sufficient here, but condition (2.2.18) will be convenient for later purpose.

However, dealing with random functions one should be aware of some pitfalls. First, if $\underline{f}(\theta)$ is a random function on an uncountable subset Θ of a Euclidean space, then $\sup_{\theta\in\Theta} \underline{f}(\theta)$ and $\inf_{\theta\in\Theta} \underline{f}(\theta)$ are not automatically random variables, for:

$$\{\omega\in\Omega : \inf_{\theta\in\Theta} f(\theta,\omega) \leq t\} = \bigcup_{\theta\in\Theta} \{\omega\in\Omega : f(\theta,\omega) \leq t\}$$

and

$$\{\omega\in\Omega : \sup_{\theta\in\Theta} f(\theta,\omega) \leq t\} = \bigcap_{\theta\in\Theta} \{\omega\in\Omega : f(\theta,\omega) \leq t\}$$

are then uncountable unions and intersections, respectively, of members of the Borel field $\mathfrak{F}$ and therefore not necessarily members of $\mathfrak{F}$ themselves. Another pitfall is that if $\underline{\theta}$ is a random vector in an uncountable subset Θ of a Euclidean space and if $\underline{f}(\theta)$ is a random function on Θ, then $\underline{f}(\underline{\theta})$ is not necessarily a random variable, because

$$\{\omega\in\Omega : f(\theta(\omega),\omega) \leq t\} = \bigcup_{\theta_*\in\Theta} [\{\omega\in\Omega : f(\theta_*,\omega) \leq t\} \cap \{\omega\in\Omega : \theta(\omega) = \theta_*\}]$$

is an uncountable union of members of $\mathfrak{F}$.

These problems can be overcome if we assume that the random function $\underline{f}(\theta)$ is <u>separable</u> [see Gihman and Skorohod (1974), chapter III, section 2]. However, in this study we shall only deal with random functions of the type

$$\underline{f}(\theta) = \phi(\theta,\underline{x}),$$

where ϕ is a continuous real function on $\Theta \times R^m$ with Θ a subset of R^k and $\underline{x}$ is a random vector in R^m, and for this case we do not need the separability concept, because of the following easy theorem: [lemma 2 of Jennrich(1969)]

<u>Theorem 2.3.1</u>: If $\phi(\theta,x)$ is a continuous real function on $\Theta \times R^m$, where Θ is a compact subset of R^k, then $\sup_{\theta\in\Theta} \phi(\theta,x)$ and $\inf_{\theta\in\Theta} \phi(\theta,x)$ are continuous functions on R^m.

<u>Proof</u>: Choose $x_0 \in R^m$ and $\varepsilon > 0$ arbitrarily. Since Θ is compact and $\Sigma = \{x\in R^m : |x - x_0| \leq \delta^*\}$ is compact for each $\delta^* > 0$ it follows that $\phi(\theta,x)$ is uniformly continuous on the compact set $\Theta \times \Sigma$. So $\delta > 0$ can be chosen such that for all $\theta_*,\theta\in\Theta$ and all $x_*,x\in\Sigma$ satisfying $|(\theta_*,x_*) - (\theta,x)| < \delta$ we have $|\phi(\theta_*,x_*) - \phi(\theta,x)| < \varepsilon$, hence for every $\theta\in\Theta$ and $|x-x_0| < \delta$,

$$\phi(\theta,x_0) - \varepsilon \leq \phi(\theta,x) \leq \phi(\theta,x_0) + \varepsilon.$$

Consequently:

$$\sup_{\theta\in\Theta} \phi(\theta,x_0) - \varepsilon \leq \sup_{\theta\in\Theta} \phi(\theta,x) \leq \sup_{\theta\in\Theta} \phi(\theta,x_0) + \varepsilon.$$

if $|x-x_0| < \delta$. This shows that $\sup_{\theta\in\Theta} \phi(\theta,x)$ is continuous in x_0. By changing "sup" in "inf" we see that also $\inf_{\theta\in\Theta} \phi(\theta,x)$ is continuous in x_0. □

Since continuous real functions are Borel measurable [see theorem 2.1.4] it follows now that for the case under review, $\sup_{\theta\in\Theta} \phi(\theta,\underline{x})$ and $\inf_{\theta\in\Theta} \phi(\theta,\underline{x})$ are random variables if Θ is compact. But this holds also if Θ is unbounded, provided that

$$\Theta_n = \Theta \cap \{\theta\in R^k : |\theta| \leq n\}.$$

is compact for each n, because we then have

$$\sup_{\theta\in\Theta} \phi(\theta,x) = \lim_{n\to\infty} \sup_{\theta\in\Theta_n} \phi(\theta,x),$$

$$\inf_{\theta\in\Theta} \phi(\theta,x) = \lim_{n\to\infty} \inf_{\theta\in\Theta_n} \phi(\theta,x),$$

which by theorem 2.1.3 and theorem 2.3.1 are Borel measurable functions. Finally, since $\phi(\theta,x)$ is Borel measurable on $\Theta \times R^m$ it follows that for any pair of random vectors $\underline{\theta}$ and $\underline{x}$ in $\Theta \times R^m$, $\phi(\underline{\theta},\underline{x})$ is a random variable.

Let us return to more general random functions. The properties of a random function $\underline{f}(\theta)$ may differ for different ω in Ω. For two points ω_1 and ω_2 in Ω it is possible for example that $f(\theta,\omega_1)$ is continuous and $f(\theta,\omega_2)$ is discontinuous at the same θ. It is even possible that a random function is not defined for each θ in a set in $\mathcal{F}$ depending on θ with probability zero.

In this study we shall always consider properties of random functions holding almost surely, which means that a property of $\underline{f}(\theta) = f(\theta,\omega)$ holds for all ω in a set $E\in\mathcal{F}$ with $P(E) = 1$. Thus for example the statement: "$\underline{f}(\theta)$ is a.s. continuous on Θ " means that there is a null set N such that $f(\theta,\omega)$ is continuous on Θ for all $\omega\in\Omega \setminus N$.

Next we shall pay attention to convergence of random functions, as here we are confronted with a similar problem. Let $\underline{f}(\theta)$ and $\underline{f}_n(\theta)$ be random functions on a subset Θ of R^k such that for each $\theta\in\Theta$, $\underline{f}_n(\theta) \to \underline{f}(\theta)$ a.s. as $n\to\infty$. Then at first sight we should expect from definition 2.2.2 that there is a null set N and an integer function $n_0(\omega,\theta,\varepsilon)$ such that for every $\varepsilon>0$ and every $\omega\in\Omega\setminus N$, $|f_n(\theta,\omega) - f(\theta,\omega)| \leq \varepsilon$ if $n \geq n_0(\omega,\theta,\varepsilon)$. But reading definition 2.2.2

carefully we see that this is not correct, because the null set N may depend on θ: $N=N_\theta$. Then again at first sight we may reply that this does not matter because we could choose $N = \bigcup_{\theta\in\Theta} N_\theta$ as a null set. But the problem now is that we are not sure whether $N\in\mathcal{F}$, for only countable unions of members of $\mathcal{F}$ are surely members of $\mathcal{F}$ themselves. Thus although $N_\theta\in\mathcal{F}$ for each $\theta\in\Theta$, this is not necessarily the case for $\bigcup_{\theta\in\Theta} N_\theta$ if Θ is uncountable. Moreover, even if $\bigcup_{\theta\in\Theta} N_\theta\in\mathcal{F}$, it may fail to be a null set itself if Θ is uncountable. For example, let $\Theta=\Omega=[0,1]$, let P be the Lebesgue measure on [0,1] and let $N_\theta=\{\theta\}$ for $\theta\in[0,1]$. Then $P(\bigcup_{\theta\in\Theta} N_\theta)=P(\Omega)=1$, while obviously the N_θ's are null sets.

As is well known, uniform convergence of (real) nonrandom functions, for example $\phi_n(\theta)\rightarrow\phi(\theta)$ uniformly on Θ as $n\rightarrow\infty$, can be defined as

$$\sup_{\theta\in\Theta} |\phi_n(\theta) - \phi(\theta)| \rightarrow 0 \text{ for } n \rightarrow \infty.$$

Dealing with uniform a.s. convergence of random functions

$$\underline{f}_n(\theta) \rightarrow \underline{f}(\theta) \text{ a.s. uniformly on } \Theta,$$

a suitable definition is therefore:

$$\sup_{\theta\in\Theta} |\underline{f}_n(\theta) - \underline{f}(\theta)| \rightarrow 0 \text{ a.s. as } n\rightarrow\infty,$$

or in other words: there is a null set N and an integer function $n_0(\omega,\varepsilon)$ such that for every $\varepsilon > 0$ and every $\omega \in \Omega \setminus N$,

$$\sup_{\theta\in\Theta} |f_n(\theta,\omega) - f(\theta,\omega)| \leq \varepsilon \text{ if } n \geq n_0(\omega,\varepsilon).$$

However, this has only a probabilistic meaning if $\sup_{\theta\in\Theta} |\underline{f}_n(\theta) - \underline{f}(\theta)|$ is a random variable for each n. Only if so, we shall say that $\underline{f}_n(\theta) \rightarrow \underline{f}(\theta)$ a.s. uniformly on Θ. Nevertheless, if $\sup_{\theta\in\Theta} |\underline{f}_n(\theta) - \underline{f}(\theta)|$ is not a random variable but if $\underline{f}_n(\theta,\omega) \rightarrow f(\theta,\omega)$ uniformly on Θ for every ω in Ω except on a null set, then we still have a useful property, as will turn out. In this case we shall say that $\underline{f}_n(\theta) \rightarrow \underline{f}(\theta)$ a.s. pseudo-uniformly on Θ.

Definition 2.3.2: Let $\underline{f}(\theta)$ and $\underline{f}_n(\theta)$ be random functions on a subset Θ of a Euclidean space, and let $\{\Omega,\mathcal{F},P\}$ be the probability space involved. Then

(a) $\underline{f}_n(\theta) \rightarrow \underline{f}(\theta)$ a.s. pointwise on Θ if for every $\theta \in \Theta$ there is a null set N_θ in $\mathcal{F}$ and for every $\varepsilon > 0$ and every $\omega\in\Omega\setminus N_\theta$ a number $n_0(\omega,\theta,\varepsilon)$ such that $|f_n(\theta,\omega) - f(\theta,\omega)| \leq \varepsilon$ if $n \geq n_0(\omega,\theta,\varepsilon)$

(b) $\underline{f}_n(\theta) \to \underline{f}(\theta)$ a.s. uniformly on Ⓗ if

(i) $\sup_{\theta\in Ⓗ} |\underline{f}_n(\theta) - \underline{f}(\theta)|$ is a random variable for n=1,2,...

and if

(ii) there is a null set N and for every $\varepsilon>0$, and $\omega\in\Omega\setminus N$ a number $n_0(\omega,\varepsilon)$ such that

$\sup_{\theta\in Ⓗ} |f_n(\theta,\omega) - f(\theta,\omega)| \leq \varepsilon$ if $n \geq n_0(\omega,\varepsilon)$.

(c) $\underline{f}_n(\theta) \to \underline{f}(\theta)$ a.s. pseudo-uniformly on Ⓗ if condition (ii) in (b) holds, but not necessarily condition (i).

Similar as in the case of a.s. uniform convergence of random functions the uniform convergence in probability of $\underline{f}_n(\theta)$ to $\underline{f}(\theta)$ can be defined by $\text{plim} \sup_{\theta\in Ⓗ} |\underline{f}_n(\theta) - \underline{f}(\theta)| = 0$, provided that $\sup_{\theta\in Ⓗ} |\underline{f}_n(\theta) - \underline{f}(\theta)|$ is a random variable for n=1,2,... . In that case it follows from theorem 2.2.4 that $\underline{f}_n(\theta) \to \underline{f}(\theta)$ in pr. uniformly on Ⓗ if and only if every subsequence (n_k) of (n) contains a further subsequence (n_{k_j}) such that $\underline{f}_{n_{k_j}}(\theta) \to \underline{f}(\theta)$ a.s. uniformly on Ⓗ . This suggests how to define pseudo-uniform convergence in pr.:

> <u>Definition 2.3.3</u>: Let $\underline{f}_n(\theta)$ and $\underline{f}(\theta)$ be random functions on a subset Ⓗ of a Euclidean space. Then
>
> a) $\underline{f}_n(\theta) \to \underline{f}(\theta)$ in pr. uniformly on Ⓗ if $\sup_{\theta\in Ⓗ} |\underline{f}_n(\theta) - \underline{f}(\theta)|$ is a random variable for n=1,2,... satisfying
>
> $\text{plim} \sup_{\theta\in Ⓗ} |\underline{f}_n(\theta) - \underline{f}(\theta)| = 0$
>
> b) $\underline{f}_n(\theta) - \underline{f}(\theta)$ in pr. pseudo-uniformly in Ⓗ if every subsequence (n_k) of (n) contains a further subsequence (n_{k_j}) such that $\underline{f}_{n_{k_j}}(\theta) \to \underline{f}(\theta)$ a.s. pseudo-uniformly on Ⓗ .

<u>Remark</u>. In this study we shall often conclude

$\sup_{\theta\in Ⓗ} |\underline{f}_n(\theta) - \underline{f}(\theta)| \to 0$ a.s.

or

$\text{plim} \sup_{\theta\in Ⓗ} |\underline{f}_n(\theta) - \underline{f}(\theta)| = 0$

instead of

$\underline{f}_n(\theta) \to f(\theta)$ a.s. uniformly on Ⓗ

or

$\underline{f}_n(\theta) \to \underline{f}(\theta)$ in pr. uniformly on Ⓗ ,

respectively. In these cases it will be clear from the context that $\sup_{\theta\in Ⓗ} |\underline{f}_n(\theta) - \underline{f}(\theta)|$ is a random variable for n=1,2,...... .

We are now able to generalize theorem 2.2.5 for random functions.

Theorem 2.3.2. Let $(\underline{f}_n(\theta))$ be a sequence of random functions on a Borel subset $\textcircled{H}$ of a Euclidean space. Let $\underline{f}(\theta)$ be an a.s. continuous random function on $\textcircled{H}$. Let $\underline{x}_n$ and $\underline{x}$ be random vectors in $\textcircled{H}$ such that $P(\underline{x}\in \textcircled{H})=1$ and $P(\underline{x}_n\in \textcircled{H})=1$ for n=1,2,... . Suppose that $\underline{f}(\underline{x})$ is a random variable and that $\underline{f}_n(\underline{x}_n)$ is a random variable for n=1,2,... . If:

a) $\underline{x}_n\to\underline{x}$ a.s. and $\underline{f}_n(\theta)\to\underline{f}(\theta)$ a.s. pseudo-uniformly on $\textcircled{H}$.

or if:

b) $\underline{x}_n\to\underline{x}$ in pr. and $\underline{f}_n(\theta)-\underline{f}(\theta)$ in pr. pseudo-uniformly on $\textcircled{H}$,

then

a) $\underline{f}_n(\underline{x}_n)\to\underline{f}(\underline{x})$ a.s.

or

b) $\underline{f}_n(\underline{x}_n)\to\underline{f}(\underline{x})$ in pr.,

respectively.

Proof: a) Let $(\Omega,\mathfrak{F},P)$ be the probability space. Let N_1 be the null set on which $x_n(\omega)\to x(\omega)$ fails to hold, let N_2 be the null set on which $f(\theta,\omega)$ fails to be continuous, let N_3 and $N_{3,n}$ be null sets on which $x(\omega)\in \textcircled{H}$ and $x_n(\omega)\in \textcircled{H}$, respectively, fail to hold and finally let N_4 be the null set on which $\sup_{\theta\in \textcircled{H}} |f_n(\theta,\omega)- f(\theta,\omega)|\to 0$ fails to hold. Put

$$N = N_1 \cup N_2 \cup N_3 \cup \{\bigcup_{n=1}^{\infty} N_{3,n}\} \cup N_4$$

Then $N\in\mathfrak{F}$, $P(N)=0$ and for $\omega\in\Omega\setminus N$ we have

$$|f_n(x_n(\omega),\omega)- f(x(\omega),\omega)|\leq|f_n(x_n(\omega),\omega)- f(x_n(\omega),\omega)|+|f(x_n(\omega),\omega)- f(x(\omega),\omega)|\leq$$

$$\leq \sup_{\theta\in \textcircled{H}} |f_n(\theta,\omega)- f(\theta,\omega)|+|f(x_n(\omega),\omega)- f(x(\omega),\omega)|\to 0 \quad \text{as} \quad n\to\infty.$$

This proves part a). Part b) follows from a) by using theorem 2.2.4.

2.3.2 Uniform strong and weak laws of large numbers

Next we shall extend the theorems 1 and 2 of Jennrich (1969). We shall closely follow Jennrich's proof, but instead of the Helly - Bray theorem (theorem 2.2.10) we shall now use theorem 2.2.17. The extension involved is:

Theorem 2.3.3. Let $\underline{x}_1,\underline{x}_2,\ldots.$ be a sequence of independent random vectors in R^k with distribution functions $F_1,F_2,\ldots.$, respectively.

Let $f(x,\theta)$ be a continuous real function on $R^k \times \textcircled{H}$, where $\textcircled{H}$ is a compact subset of R^m.

Let $\psi(a)= \sup_{\{|x|\leq a\}} \sup_{\theta\in \textcircled{H}} |f(x,\theta)|$

If:

$$\frac{1}{n}\sum_{j=1}^{n} F_j \to G \text{ properly} \tag{2.3.1}$$

and both:

$$\sup_n \frac{1}{n}\sum_{j=1}^{n} E\psi(|\underline{x}_j|)^{1+\delta} < \infty \text{ for some } \delta > 0, \tag{2.3.2}$$

$$E(\psi(|\underline{x}_j|))^2 = O(j^{\mu}) \text{ for some } \mu<1, \tag{2.3.3}$$

then $\frac{1}{n}\sum_{j=1}^{n} f(\underline{x}_j,\theta) \to \int f(x,\theta)dG(x)$ a.s. uniformly on Ⓗ.

Proof: From theorem 2.3.1 it follows that $\sup_{\theta\in Ⓗ}|f(x,\theta)|$ is a continuous function of x and consequently it follows from lemma 2.2.1 that $\psi(a)$ is a continuous function on $[0,\infty)$. Hence $\psi(|\underline{x}_j|)$ is a random variable, and therefore the conditions (2.3.2) and (2.3.3) make sense.

For the sake of convenience and clearity we shall label the main steps of the proof.

Step 1:

Choose θ_0 arbitrarily in Ⓗ and put $\Gamma_\delta = \{\theta\in R^m : |\theta - \theta_0| \leq \delta\} \cap Ⓗ$, for $\delta \geq 0$. Then for any $\delta \geq 0$, $\sup_{\theta\in\Gamma_\delta} f(x,\theta)$ and $\inf_{\theta\in\Gamma_\delta} f(x,\theta)$ are continuous functions on R^k, because Γ_δ is a closed subset of a compact set and therefore [see Rudin (1976), theorem 2.35] compact itself (compare theorem 2.3.1). Moreover,

$$|\sup_{\theta\in\Gamma_\delta} f(x,\theta)| \leq \sup_{\theta\in Ⓗ} |f(x,\theta)| \leq \psi(|x|) \tag{2.3.4}$$

$$|\inf_{\theta\in\Gamma_\delta} f(x,\theta)| \leq \sup_{\theta\in Ⓗ} |f(x,\theta)| \leq \psi(|x|) . \tag{2.3.5}$$

Thus it follows from theorem 2.2.17 and the conditions (2.3.1), (2.3.2) and (2.3.3) that

$$\frac{1}{n}\sum_{j=1}^{n} \sup_{\theta\in\Gamma_\delta} f(\underline{x}_j,\theta) \to \int \sup_{\theta\in\Gamma_\delta} f(x,\theta)dG(x) \to \text{ a.s.} \tag{2.3.6}$$

and

$$\frac{1}{n}\sum_{j=1}^{n} \inf_{\theta\in\Gamma_\delta} f(\underline{x}_j,\theta) \to \int \inf_{\theta\in\Gamma_\delta} f(x,\theta)dG(x) \text{ a.s. .} \tag{2.3.7}$$

Step 2:

Similar as in the proof of theorem 2.2.5 it follows from the conditions (2.3.1) and (2.3.2) that:

$$\int\psi(|x|)dG(x)<\infty ,$$

while by (2.3.4) and (2.3.5) we have for every $\delta \geq 0$:

$$|\sup_{\theta\in\Gamma_\delta} f(x,\theta)- \inf_{\theta\in\Gamma_\delta} f(x,\theta)|\leq 2\psi(|x|) .$$

Therefore

$$\lim_{\delta\downarrow 0}|\int \sup_{\theta\in\Gamma_\delta} f(x,\theta)dG(x)- \int\inf_{\theta\in\Gamma_\delta} f(x,\theta)dG(x)|=0 \qquad (2.3.8)$$

by bounded convergence.

Step 3:

Choose $\varepsilon>0$ arbitrarily. From (2.3.8) it follows that $\delta>0$ can be chosen so small, say $\delta=\delta(\varepsilon)$, that:

$$0\leq\int\sup_{\theta\in\Gamma_{\delta}(\varepsilon)} f(x,\theta)dG(x)- \int\inf_{\theta\in\Gamma_{\delta}(\varepsilon)} f(x,\theta)dG(x)\leq\tfrac{1}{2}\varepsilon . \qquad (2.3.9)$$

Let $\{\Omega,\mathfrak{F},P\}$ be the probability space involved. From (2.3.6) and (2.3.7) it follows that there is a null set N and for each $\omega\in\Omega\setminus N$ a number $M_0(\omega,\varepsilon)$ such that:

$$|\tfrac{1}{n}\textstyle\sum_{j=1}^{n} \sup_{\theta\in\Gamma_{\delta}(\varepsilon)} f(x_j(\omega),\theta)- \int\sup_{\theta\in\Gamma_{\delta}(\varepsilon)} f(x,\theta)dG(x)|\leq\tfrac{1}{2}\varepsilon , \qquad (2.3.10)$$

$$|\tfrac{1}{n}\textstyle\sum_{j=1}^{n} \inf_{\theta\in\Gamma_{\delta}(\varepsilon)} f(x_j(\omega),\theta)- \int\inf_{\theta\in\Gamma_{\delta}(\varepsilon)} f(x,\theta)dG(x)|\leq\tfrac{1}{2}\varepsilon \qquad (2.3.11)$$

if $n\geq n_0(\omega,\varepsilon)$. From (2.3.9), (2.3.10) and (2.3.11) it follows now that for all $\omega\in\Omega\setminus N$, all $n\geq n_0(\omega,\varepsilon)$ and all $\theta\in\Gamma_{\delta(\varepsilon)}$:

$$\tfrac{1}{n}\textstyle\sum_{j=1}^{n} f(x_j(\omega),\theta)- \int f(x,\theta)dG(x)\leq \tfrac{1}{n}\sum_{j=1}^{n} \sup_{\theta\in\Gamma_{\delta(\varepsilon)}} f(x_j(\omega),\theta)- \int_{\theta\in\Gamma_{\delta(\varepsilon)}} f(x,\theta)dG(x)\leq$$

$$\leq|\tfrac{1}{n}\textstyle\sum_{j=1}^{n} \sup_{\theta\in\Gamma_{\delta}(\varepsilon)} f(x_j(\omega),\theta)- \int\sup_{\theta\in\Gamma_{\delta(\varepsilon)}} f(x,\theta)dG(x)|+$$

$$+|\int\sup_{\theta\in\Gamma_{\delta(\varepsilon)}} f(x,\theta)dG(x)- \int\inf_{\theta\in\Gamma_{\delta(\varepsilon)}} f(x,\theta)dG(x)|\leq\varepsilon$$

and similarly:

$$\tfrac{1}{n}\textstyle\sum_{j=1}^{n} f(x_j(\omega),\theta)- \int f(x,\theta)dG(x)\geq -\varepsilon.$$

Thus for $\omega\in\Omega\setminus N$ and $n\geq n_0(\omega,\varepsilon)$ we have:

$$\sup_{\theta\in\Gamma_{\delta(\varepsilon)}} \left|\frac{1}{n}\sum_{j=1}^{n} f(x_j(\omega),\theta) - \int f(x,\theta)dG(x)\right| \leq \varepsilon.$$

We note that the null set N and the number $n_0(\omega,\varepsilon)$ depend on the set $\Gamma_{\delta(\varepsilon)}$, which in its turn depends on θ_0 and ε. Thus the above result should be restated as follows. For every θ_0 in Ⓗ and every $\varepsilon > 0$ there is a null set $N(\theta_0,\varepsilon)$ and an integer function $n_0(.,\varepsilon,\theta_0)$ on $\Omega\setminus N(\theta_0,\varepsilon)$ such that for $\omega\in\Omega\setminus N(\theta_0,\varepsilon)$ and $n\geq n_0(\omega,\varepsilon,\theta_0)$:

$$\sup_{\theta\in\Gamma_{\delta(\varepsilon)}(\theta_0)} \left|\frac{1}{n}\sum_{j=1}^{n} f(x_j(\omega),\theta) - \int f(x,\theta)dG(x)\right| \leq \varepsilon, \tag{2.3.12}$$

where $\delta(\varepsilon)$ is some positive (real) valued function of $\varepsilon\in(0,\infty)$ and

$$\Gamma_\delta(\theta_0) = \{\theta\in R^k\colon |\theta - \theta_0|\leq\delta\}\cap Ⓗ . \tag{2.3.13}$$

<u>Step 4</u>:

The collection of sets $\{\theta\in R^k\colon|\theta-\theta_0|<\delta\}$ with $\theta_0\in$Ⓗ is an open covering of Ⓗ. Since Ⓗ is compact, there exists by definition of compactness an finite covering. Thus there are a finite number of points in Ⓗ, say $\theta_{\delta,1},\ldots,\theta_{\delta,r_\delta}$ with $r_\delta < \infty$ such that

$$Ⓗ \subset \bigcup_{i=1}^{r_\delta}\{\theta\in R^k\colon|\theta - \theta_{\delta,i}|<\delta\} \subset \bigcup_{i=1}^{r_\delta}\{\theta\in R^k|\theta - \theta_{\delta,i}|< \delta\}.$$

From (2.3.13) we therefore have:

$$Ⓗ = \bigcup_{i=1}^{r_\delta}\Gamma_\delta(\theta_{\delta,i}) . \tag{2.3.14}$$

Now put:

$$N_\varepsilon = \bigcup_{i=1}^{r_{\delta(\varepsilon)}} N(\theta_{\delta(\varepsilon),i},\varepsilon),$$

$$n_*(\omega,\varepsilon) = \max_{1\leq i\leq r_{\delta(\varepsilon)}} n_0(\omega,\varepsilon,\theta_{\delta(\varepsilon),i}).$$

Then by (2.3.13) and (2.3.14) we have for $\omega\in\Omega\setminus N_\varepsilon$ and $n\geq n_*(\omega,\varepsilon)$,

$$\sup_{\theta\in Ⓗ} \left|\frac{1}{n}\sum_{j=1}^{n} f(x_j(\omega),\theta) - \int f(x,\theta)dG(x)\right| \leq$$

$$\leq \max_{1\leq i\leq r_{\delta(\varepsilon)}}\ \sup_{\theta\in\Gamma_{\delta(\varepsilon)}(\theta_{\delta(\varepsilon),i})} \left|\frac{1}{n}\sum_{j=1}^{n} f(x_j(\omega),\theta) - \int f(x,\theta)dG(x)\right| \leq \varepsilon.$$

Since, similar as in the proof of theorem 2.2.3, it can be shown that the null set N_ε can be chosen independent of ε, it follows now that

$$\frac{1}{n}\sum_{j=1}^{n} f(\underline{x}_j,\theta) \to \int f(x,\theta)dG(x) \text{ a.s. pseudo-uniformly on } \textcircled{H} . \qquad (2.3.15)$$

Step 5:

From (2.3.8) it follows that $\int f(x,\theta)dG(x)$ is a continuous function on $\textcircled{H}$. Using theorem 2.3.1, it is now easy to verify that $\sup_{\theta\in\textcircled{H}} |\frac{1}{n}\sum_{j=1}^{n} f(\underline{x}_j,\theta) - \int f(x,\theta)dG(x)|$ is a random variable, so that (2.3.15) becomes

$$\frac{1}{n}\sum_{j=1}^{n} f(\underline{x}_j,\theta) \to \int f(x,\theta)dG(x)| \text{ a.s. uniformly on } \textcircled{H} .$$

This completes the proof. □

If condition (2.3.3) is not satisfied then we can no longer apply theorem 2.2.17 for proving (2.3.6) and (2.3.7). But applying theorem 2.2.16 we see that (2.3.6) and (2.3.7) still hold in probability. From theorem 2.2.4 it then follows that any subsequence (n_k) of (n) contains further subsequences $(n_{k_m}^{(1)})$ and $(n_{k_m}^{(2)})$, say such that

$$\frac{1}{n_{k_m}} \sum_{j=1}^{n_{k_m}^{(1)}} \sup_{\theta\in\Gamma_\delta} f(\underline{x}_j,\theta) \to \int \sup_{\theta\in\Gamma_\delta} f(x,\theta)dG(x) \quad \text{a.s. as } m\to\infty ,$$

$$\frac{1}{n_{k_m}} \sum_{j=1}^{n_{k_m}^{(2)}} \inf_{\theta\in\Gamma_\delta} f(\underline{x}_j,\theta) \to \int \inf_{\theta\in\Gamma_\delta} f(x,\theta)dG(x) \quad \text{a.s. as } m\to\infty .$$

But $(n_{k_m}^{(2)})$ may also be considered as a subsequence of $(n_k^{(1)})$, hence without loss of generality we may assume that these further subsequences are equal:

$$n_{k_m} = n_{k_m}^{(1)} = n_{k_m}^{(2)} .$$

We now conclude from the argument in the proof of theorem 2.3.3 that:

$$\sup_{\theta\in\textcircled{H}} |\frac{1}{n_{k_m}} \sum_{j=1}^{n_{k_m}} f(\underline{x}_j,\theta) - \int f(x,\theta)dG(x)| \to 0 \text{ a.s. as } m\to\infty .$$

Again using theorem 2.2.4 we then conclude:

Theorem 2.3.4. Let the conditions of theorem 2.3.3 be satisfied, except (2.3.3). Then

$$\frac{1}{n}\sum_{j=1}^{n} f(\underline{x}_j,\theta) \to \int f(x,\theta)dG(x) \text{ in pr. uniformly on } \textcircled{H} .$$

Functions ψ as considered in theorem 2.3.3 will be frequently used in this study. Therefore a more compact notation is convenient [Compare definition 2.2.6].

Definition 2.3.4. Let $\phi(x,\theta)$ be a real function on X x Θ, where X is a Euclidean space. Then by $\overline{\phi(x,\theta)}^{\Theta}$ we denote:

$$\overline{\phi(x,\theta)}^{\Theta} = \psi(|x|),$$

where $\psi(a) = \sup_{\{|x|\leq a\}} \sup_{\theta\in\Theta} |\phi(x,\theta)|, \; a \geq 0 .$

The reader is invited to verify the following easy but useful (in)equalities.

$$\overline{\{\phi(x,\theta)^p\}}^{\Theta} = \{\overline{\phi(x,\theta)}^{\Theta}\}^p \text{ for } p > 0 , \tag{2.3.16}$$

$$\overline{\phi_1(x,\theta)\phi_2(x,\theta)}^{\Theta} \leq \{\overline{\phi_1(x,\theta)}^{\Theta}\}\{\overline{\phi_2(x,\theta)}^{\Theta}\} , \tag{2.3.17}$$

$$\overline{\phi_1(x,\theta)+\phi_2(x,\theta)}^{\Theta} \leq \overline{\phi_1(x,\theta)}^{\Theta} + \overline{\phi_2(x,\theta)}^{\Theta} . \tag{2.3.18}$$

Moreover, it follows from lemma 2.2.1 and theorem 2.3.1 that if $\phi(x,\theta)$ is continuous on X x Θ, where X is an Euclidean space and Θ a compact subset of an Euclidean space, then $\overline{\phi(x,\theta)}^{\Theta}$ is a continuous function of $|x|$.

2.4 Characteristic functions, stable distributions and a central limit theorem

Characteristic functions and central limit theorems are standard tools in mathematical statistics and hence in econometrics. Therefore we only summarize the for our purpose important results, and for the proofs we refer to textbooks like Feller (1966), Chung (1974), or Wilks (1963).

Definition 2.4.1: Let $\underline{x}$ be a random vector in R^k with distribution function F. The characteristic function of this distribution is the following complex valued function $\phi(t)$ on R^k:

$$\phi(t) = Ee^{i\,t'\underline{x}} = E\cos(t'\underline{x}) + iE\sin(t'\underline{x}) .$$

Distributions are fully determined by their characteristic functions: distributions are equal if and only if their characteristic functions are equal. Moreover, convergence in distribution is closely related with convergence of characteristic functions:

<u>Theorem 2.4.1</u>. Let (F_n) be a sequence of distribution functions on R^k and let (ϕ_n) be the sequence of corresponding characteristic functions. If $F_n \to F$ properly then $\phi_n(t) \to \int e^{i\,t'x}\, dF(x)$ pointwise on R^k.

If $\phi_n(t) \to \phi(t)$ pointwise on R^k and $\phi(t)$ is continuous at $t=0$ then there exists a unique distribution function F such that $\phi(t)=\int e^{i\,t'x} dF(x)$ and $F_n \to F$ properly.

We recall that the characteristic function of the k-variate normal distribution $N_k(\mu,\Sigma)$ is given by $e^{i\,\mu't - \frac{1}{2}\,t'\Sigma t}$ (See Anderson (1958)).Now consider a random vector $\underline{x}$ in R^k with mathematical expectation*) $E\underline{x}=\mu$ and variance matrix $E(\underline{x}-\mu)(\underline{x}-\mu)'=\Sigma$.**) Suppose that for every vector ξ in R^k, $\xi'\underline{x}$ is normally distributed; we may then conclude that $\xi'\underline{x} \sim N(\xi'\mu,\xi'\Sigma\xi)$. Then we have for every t in R and every $\xi \in R^k$

$$Ee^{i\,t\,\xi'\underline{x}} = e^{i\,t\,\mu'\xi - \frac{1}{2}t^2\xi'\Sigma\xi},$$

hence, substituting t=1,

$$Ee^{i\xi'\underline{x}} = e^{i\,\mu'\xi - \frac{1}{2}\xi'\Sigma\xi}$$ for all $\xi \in R^k$. So we conclude that $\underline{x} \sim N(\mu,\Sigma)$.

Consequently we have:

<u>Theorem 2.4.2</u>. Let $(\underline{x}_n)$ be a sequence of random vectors in R^k. If for all vectors $\xi \in R^k$, $\xi'\underline{x}_n$ converges in distribution to a normal distribution $N(\mu'\xi,\xi'\Sigma\xi)$, where Σ is a positive (semi)definite matrix, then $\underline{x}_n$ converges in distribution to the k-variate normal distribution $N_k(\mu,\Sigma)$.

This theorem is very useful in combination with the central limit theorem (theorem 2.4.6 below).

A random variable or random vector $\underline{x}$ is symmetrically distributed if $\underline{x}$ and $-\underline{x}$ have the same distribution. Since this is only the case if the characteristic functions are equal we have:

---

*) The mathematical expectation of a vector or matrix of random variables is defined as the vector or matrix, respectively, of their mathematical expectations. The variance of a random vector $\underline{x}$ is the matrix $E(\underline{x} - E\underline{x})(\underline{x} - E\underline{x})'$.

**) If Σ is a singular matrix then the normal distribution involved is said to be singular.

$$Ee^{it'x} = E\cos(t'\underline{x}) + i\,E\sin(t'\underline{x}) = Ee^{-it'\underline{x}} = E\cos(t'\underline{x}) - i\,E\sin(t'\underline{x}) ,$$

hence $E\sin(t'\underline{x})=0$ for all t. Thus:

> Theorem 2.4.3. A distribution is symmetric if and only if the characteristic function involved is real valued everywhere.

This simple theorem will be the starting point for our theory about estimating nonlinear structural equations.

A distribution function F on R^k is called absolutely continuous if there exists a non-negative function f on R^k satisfying:

$$F(t) = \int_{\{u \leq t\}} f(u)du$$

(where u and t are vectors in R^k). The function f is called a density*) of the distribution. For such distibutions we have the following results.

> Theorem 2.4.4. A distribution function F is continuously differentiable up to the k-th order if its characteristic function $\phi(t)$ satisfies:
>
> $$\int_{-\infty}^{+\infty} |t|^{k-1} |\phi(t)|\, dt < \infty .$$

This can be proved by generalizing the following theorem to derivatives, noting that by the condition $\int_{-\infty}^{+\infty} |t|^{k-1}|\phi(t)|dt < \infty$ we may differentiate k-times under the integral below.

> Theorem 2.4.5. Let F be a distribution function on R^k and let ϕ be its characteristic function. If ϕ is absolutely integrable then F is absolutely continuous and its density f satisfies:
>
> $$f(x) = \left(\frac{1}{2\pi}\right)^k \int_{R^k} e^{-it'x}\phi(t)dt =$$
>
> $$\left(\frac{1}{2\pi}\right)^k \int_{R^k} \cos(t'x)\mathrm{Re}\phi(t)dt + \left(\frac{1}{2\pi}\right)^k \int_{R^k} \sin(t'x)\mathrm{Im}\phi(t)dt ,$$

*) Which is unique, possibly except on a set with zero Lebesgue measure.

where "Re" and "Im" stand for Real part and Imaginary part, respectively, of the function involved.

In section 1.1 we have already mentioned the class of stable distributions. Here we shall give a brief review of the most important properties of stable distributions. For more details we refer to Feller (1966). These distributions are called "stable" because if $\underline{u}_1,...,\underline{u}_n$ are independent random drawings from a stable distribution then there are constants a_n, b_n such that $\sum_{j=1}^{n} \underline{u}_j$ and $a_n\underline{u}_1 + b_n$ are equally distributed. Thus convolutions of stable distributions are stable themselves.

The constants a_n can only be

$$a_n = n^{\frac{1}{\alpha}}, \text{ where } \alpha \in (0,2] \tag{2.4.1}$$

(see Feller (1966), theorem 1 at page 166), while the constants b_n can only be such that

$$b_n = \begin{cases} b(n^{\frac{1}{\alpha}} - n) & \text{if } \alpha \neq 1 \\ b(n \log n) & \text{if } \alpha = 1, \end{cases} \tag{2.4.2}$$

where b is some constant, independent of n (see Feller (1966), footnote 2 at page 167 and theorem 3 at page 168).

The constant α in (2.4.1) is called the <u>characteristic exponent</u>, because the characteristic function of a stable distribution is dominated by $\exp(c|t|^{\alpha})$, where c is some positive constant, while in the symmetric case the characteristic function is just $\exp(c|t|^{\alpha})$.

A distribution F is said to belong to the <u>domain of attraction</u> of a distribution G if there are sequences (c_n) and (d_n) of constants such that for any independent sample $\underline{x}_1,...,\underline{x}_n$ from F

$$P\left(\frac{\sum_{j=1}^{n} \underline{x}_j - d_n}{c_n} \leq x\right) \to G(x) \text{ properly.}$$

Such a distribution G can either be degenerate or stable. Thus a non-degenerate distribution having a domain of attraction is necessarily stable (see Feller (1966), theorems 1 and 1a at page 544). Moreover, a stable distribution with characteristic exponent α belongs to its own domain of attraction.

A necessary condition of a distribution F belonging to the domain of attraction of a stable distribution with characteristic exponent $\alpha \in (0,2]$ is that the function

$$L(x)= x^{\alpha-2} \int_{-x}^{x} y^2 dF(y) \tag{2.4.3}$$

is slowly varying:

$$\lim_{t\to\infty} \frac{L(tx)}{L(t)} = 1 \tag{2.4.4}$$

for fixed $x > 0$. If $\int_{-\infty}^{+\infty} y^2 dF(y) < \infty$ then (2.4.4) can only be true for $\alpha = 2$, hence a stable distribution with characteristic exponent $\alpha < 2$ must satisfy

$$\int_{-\infty}^{+\infty} y^2 dF(y)=\infty$$

because stable distributions belong to their own domain of attraction. Thus stable distribution with characteristic exponent $\alpha < 2$ have an infinite second moment. However, they possess absolute moments of every order $\beta < \alpha$:

$$\int_{-\infty}^{+\infty} |y|^{\beta} \, dF(y)<\infty \quad \text{for } \beta\in(0,\alpha)$$

(see Lemma 2 at page 545 of Feller (1966)).

We end up this section with Liapounov's central limit theorem for independent random variables.

<u>Theorem 2.4.6</u>. Let $\underline{s}_n=\sum_{j=1}^{k_n}\underline{x}_{n,j}$, where for each n the r.v.'s $\underline{x}_{n,1},\ldots,\underline{x}_{n,k_n}$ are independent and $k_n\to\infty$. Put $E\underline{x}_{n,j}=\alpha_{n,j}$, $\alpha_n=\sum_{j=1}^{k_n}\alpha_{n,j}$

$\sigma^2(\underline{x}_{n,j})=E(\underline{x}_{n,j}-\alpha_{n,j})^2=\sigma^2_{n,j}$, $\sigma^2_n=\sum_{j=1}^{k_n}\sigma^2_{n,j}$, assuming $\sigma_n^2 < \infty$ *).

If for some $\delta > 0$,

$$\lim_{n\to\infty} \sum_{j=1}^{k_n} E\left|\frac{\underline{x}_{n,j} - \alpha_{n,j}}{\sigma_n}\right|^{2+\delta}=0$$

then $\dfrac{\underline{s}_n - \alpha_n}{\sigma_n} \to N(0,1)$ in distr.

---

*) But <u>not</u> necessarily $\limsup \sigma_n^2 < \infty$.

2.5 Unimodal distributions

In the forthcoming sections 3.2 and 3.3 it will be shown that if the error distribution is symmetric unimodal, a consistent estimation method can be based on the property that convolutions of symmetric unimodal distributions are themselves symmetric unimodal. But the theory about unimodal distributions is scarce, and what is available is not treated systematically in standard textbooks (Feller (1966) only devoted some footnotes and excercises to it) or even completely absent, even in an otherwise excellent textbook like Chung (1974). In this section we shall therefore briefly review the (for our purpose) most important results about unimodal distributions and in particular we shall focus on the symmetric case. Our starting point is the following result due to Khintchine:

> Theorem 2.5.1. A distribution $F(u)$ is unimodal with mode at the origin if and only if it is the distribution of the product $\underline{x}=\underline{y}.\underline{z}$ of two independent random variables $\underline{y}$ and $\underline{z}$ such that $\underline{y}$ is distributed uniformly in $[0,1]$.

Proof: See Feller(1966, p.155-156).

Now let $\phi(t)$ and $\psi(t)$ be the characteristic functions of the random variables $\underline{x}$ and $\underline{z}$ above. Then :

$$\phi(t)=Ee^{it\underline{x}}=Ee^{it\underline{y}\underline{z}}=E\psi(t\underline{y})=\int_0^1 \psi(ty)dy = \frac{1}{t}\int_0^t \psi(y)dy \quad \text{if } t \neq 0 .$$

Thus $\phi(t)$ is differentiable for $t \neq 0$:

$$\phi'(t)=-\frac{1}{t^2}\int_0^t \psi(y)dy + \frac{1}{t}\psi(t) = -\frac{1}{t}\phi(t)+\frac{1}{t}\psi(t)$$

hence :

$$\psi(t)= \phi(t) + t\phi'(t) \quad \text{for } t \neq 0$$

So we have shown the "only if" part of the following theorem.(The proof of the "if" part is not hard and therefore left to the reader).

> Theorem 2.5.2. Let $\phi(t)$ be a characteristic function. Then $\phi(t)$ belongs to a unimodal distribution with mode at the origin if and only if $\phi'(t)$ exists for $t \neq 0$ and there exists a characteristic function ψ such that $\psi(t)=\phi(t)+t\phi'(t)$ for $t \neq 0$.

These theorems can be used for proving that symmetric stable distributions are unimodal. Doob writes in the appendix of Gnedenko and Kolmogorov (1954): "It is not even known whether all stable laws are unimodal. The only result known in this direction seems to be those of Wintner which state that the composition of two symmetrical unimodal distribution functions is symmetrical unimodal, and consequently that all symmetrical stable laws are unimodal". Luckacs (1970, p. 158-161) gives a rather difficult proof of the latter proposition. Our proof below is much easier.

Let $\underline{y}_1,\ldots,\underline{y}_n$ be a sample from the uniform $[0,1]$ distribution and let $\underline{z}_1,\ldots,\underline{z}_n$ be a sample from the symmetric stable distribution with characteristic function $e^{-|t|^{\alpha}}$, $\alpha\in(0,2]$, with $y_1,z_1,\ldots,\underline{y}_n,z_n$ independent. Then theorem 2.5.1 says that the distribution of $\underline{y}_j\underline{z}_j$ is unimodal, while obviously this distribution is also symmetric. Since:

> Theorem 2.5.3. Convolutions of symmetric unimodal distributions are symmetric unimodal themselves,

[See Feller (1966, exercise 21 at page 164)], it follows that the distribution of $\sum_{j=1}^{n}\underline{y}_j\underline{z}_j$ is symmetric unimodal, hence (taking conditional expectations):

$$\phi_n(t)=Ee^{it\sum_{j=1}^{n}\underline{y}_j\underline{z}_j}=E\{E[e^{it\sum_{j=1}^{n}\underline{y}_j\underline{z}_j}|\underline{y}_1,\ldots,\underline{y}_n]\}$$

$$=E\prod_{j=1}^{n}E[e^{it\underline{y}_j\underline{z}_j}|\underline{y}_j]=E\prod_{j=1}^{n}e^{-|t|^{\alpha}|\underline{y}_j|^{\alpha}}=Ee^{-|t|^{\alpha}\sum_{j=1}^{n}|\underline{y}_j|^{\alpha}}$$

is a characteristic function of a symmetric unimodal distribution, and so is

$$\phi_n((\tfrac{1}{n})^{\frac{1}{\alpha}}t)=Ee^{-|t|^{\alpha}\frac{1}{n}\sum_{j=1}^{n}|\underline{y}_j|^{\alpha}}$$

Since $\text{plim}\,\frac{1}{n}\sum_{j=1}^{n}|\underline{y}_j|^{\alpha}=\lim\frac{1}{n}\sum_{j=1}^{n}E|\underline{y}_j|^{\alpha}=\int_0^1 y^{\alpha}dy=\frac{1}{1+\alpha}$, it follows that

$$\text{plim}\,e^{-|t|^{\alpha}\frac{1}{n}\sum_{j=1}^{n}|\underline{y}_j|^{\alpha}}=e^{-\frac{1}{1+\alpha}|t|^{\alpha}}$$

and consequently by the bounded convergence theorem

$$\lim\phi_n((\tfrac{1}{n})^{\frac{1}{\alpha}}t)=e^{-\frac{1}{1+\alpha}|t|^{\alpha}}.$$

Similarly we have for $t \neq 0$

$$\lim \phi_n((\frac{1}{n})^{\frac{1}{\alpha}} t) + t \frac{d}{dt} \phi_n((\frac{1}{n})^{\frac{1}{\alpha}} t) = (1 - \frac{\alpha}{1+\alpha}|t|^\alpha) e^{-\frac{1}{1+\alpha}|t|^\alpha} , \qquad (2.5.1)$$

where obviously the right member is continuous at $t = 0$.

From theorem 2.5.2 it follows that the left member of (2.5.1) is the limit of a sequence of characteristic functions, hence by theorem 2.4.1, the right member is a characteristic function and consequently (again by theorem 2.5.2) $e^{-\frac{1}{1+\alpha}|t|^\alpha}$ is the characteristic function of a symmetric unimodal distribution

So we have proved:

Theorem 2.5.4. Symmetric stable distributions are unimodal.

3 NONLINEAR REGRESSION MODELS

In this chapter we shall consider the problem of estimating the parameters of a nonlinear regression model with additive disturbance term under a number of alternative assumptions about the distributions of errors and regressors. In section 3.1 we first discuss the least squares estimation theory of Jennrich (1969), but in a slightly more general setting than Jennrich did. Then, at the end of this section, we shall weaken some assumptions and especially we show what happens if the error distribution is symmetric stable with characteristic exponent $\alpha < 2$. In section 3.2 a nonlinear robust two-stage M-estimation method is presented. This method is especially appropriate when the second moment of the error distribution is infinite. But even when the usual assumptions about the error distribution are satisfied, but the error distribution is leptocurtic, which means that the fourth moment of this error distribution is larger than three times the square of the second moment, then robust M-estimation turns out to be asymptotically more efficient than least squares. Both least squares estimation and robust M-estimation requires some assumptions about the distributions of the regressors. In section 3.3 we therefore introduce a weighted robust two-stage M-estimation method which does not need any assumption about finiteness of moments of errors and regressors. The first-stage estimators of both robust M-estimation methods are random vector-functions of a scaling parameter. In section 3.4.1 it is shown that these vector-functions are (pseudo) uniformly consistent on a compact set of scaling parameters. Section 3.4.2 shows that the asymptotic properties of the unweighted robust M-estimator may hold even if the error distribution fails to be symmetric unimodal. In section 3.4.6 we consider some open questions concerning the choice of the function ρ. Moreover, in section 3.4.4 an easy statistic, the weighted residual kurtosis, is proposed for deciding whether least squares regression results may be improved by robust M-estimation.

3.1 Nonlinear least-squares estimation

3.1.1 Model and estimator

The asymptotic properties of nonlinear least squares estimators follow straightforwardly from maximum likelihood theory if the errors are i.i.d. normal random variables. But also if normality is not assumed, the nonlinear least squares estimator turns out to be consistent and asymptotically normally distributed under rather general conditions, as is shown by Hartley and Booker (1965), Villegas (1969), Malinvaud (1970a, 1970b) and Jennrich (1969). Especially

Jennrich's approach is an important source of inspiration for the present study, in particular his theorems 1 and 2 which show how convergence of distribution functions can be used for proving uniform convergence of means of random functions. For its own value but also as a starting point for further generalizations in the direction of M-estimators we shall therefore give now a full discussion of a more general variant of Jennrich's least squares theory. We shall do this without using his tail norm concept because this concept is a stumbling-block for these further generalizations. From the conditions of his theorems 1 and 2 it follows that Jennrich is in fact only dealing with nonlinear regression models with either bounded response functions or identically distributed regressors. However, from our theorem 2.3.3, which is a further elaboration of Jennrich's theorems 1 and 2, it follows that tail norms exist under more general conditions than Jennrich considered. Therefore we shall use directly these general conditions instead of the tail norm concept.

The nonlinear regression model to be considered has the structure:

$$\underline{y}_j = g(\underline{x}_j, \theta_0) + \underline{u}_j \ , \quad j=1,2,... \tag{3.1.1}$$

where:

Assumption 3.1.1. The $\underline{u}_j$'s are independent and identically distributed real random errors, the $\underline{x}_j$'s are independently distributed random regressors in the p-dimensional real space R^p, and are also independent of the errors, θ_0 is a parameter vector in a known compact subset Ⓗ of the q-dimensional real space R^q and the response function $g(x,\theta)$ and its partial derivatives $(\partial/\partial\theta_i)g(x,\theta)$ and $(\partial/\partial\theta_{i_1})(\partial/\partial\theta_{i_2})g(x,\theta)$ are continuous real functions on $R^p \times$ Ⓗ $,(i,i_1,i_2=1,2,...,q)$.

Of course, the $\underline{y}_j$'s and the $\underline{x}_j$'s are observable, and the $\underline{u}_j$'s are unobservable. A usual assumption about the error distribution is:

Assumption 3.1.2. The distribution of the errors, $F(u)$ say, satisfies

$\int u dF(u)=0$ and $\int |u|^{2+\delta} dF(u) < \infty$ for some $\delta > 0$. Moreover, we put $\sigma^2 = \int u^2 dF(u)$.

A nonlinear least squares estimator, $\hat{\underline{\theta}}_n$ say, of θ_0 is then a solution of the following minimalization problem:

$$\underline{Q}_n(\hat{\underline{\theta}}_n) = \inf_{\theta \in Ⓗ} \underline{Q}_n(\theta) \ ; \ \hat{\underline{\theta}}_n \in Ⓗ \ , \tag{3.1.2}$$

where

$$\underline{Q}_n(\theta)= \frac{1}{n} \sum_{j=1}^{n}(\underline{y}_j - g(\underline{x}_j,\theta))^2=\frac{1}{n}\sum_{j=1}^{n}\{\underline{u}_j+ g(\underline{x}_j,\theta_0)- g(\underline{x}_j,\theta)\}^2=$$

$$=\frac{1}{n} \sum_{j=1}^{n}\underline{u}_j^2 + 2 \frac{1}{n} \sum_{j=1}^{n}\underline{u}_j(g(\underline{x}_j, \theta_0)- g(\underline{x}_j,\theta))$$

$$+ \frac{1}{n} \sum_{j=1}^{n}(g(\underline{x}_j, \theta_0)- g(\underline{x}_j,\theta))^2. \qquad (3.1.3)$$

Our first target will be to establish further conditions such that $\hat{\underline{\theta}}_n \to \theta_0$ a.s. However, since convergence concepts as convergence a.s., in pr. or in distr. are only defined for random variables or random vectors, we have to deal with the question whether or not $\hat{\underline{\theta}}_n$ is a random vector.*) In the statistical and econometric literature this problem is often overlooked, except by Jennrich, Witting and Nölle. Jennrich gives the following lemma.

<u>Lemma 3.1.1</u>. Let Θ be a compact subset of R^m and let $f(x,\theta)$ be a real function on $R^k \times \Theta$ such that for each $\theta_0 \in \Theta$, $f(x,\theta_0)$ is Borel-measurable on R^k and for each $x_0 \in R^k$, $f(x_0,\theta)$ is continuous on Θ. Then there exists a vector function $\hat{\theta}(x)$ from R^k in to Θ with Borel measurable components, such that $f(x,\hat{\theta}(x)) = \sup_{\theta \in \Theta} f(x,\theta)$.

A similar result is proved by Witting and Nölle (1970), Hilfssatz 2.30 at page 75 and 76. Moreover, we also have the following related result, which will be used in section 3.4.1.

<u>Lemma 3.1.2</u>. Let the conditions of lemma 3.1.1 be satisfied and assume in addition that $f(x,\theta)$ is continuous on $R^k \times \Theta$. If $\hat{\theta}(x)$ is unique at $x = x_0$ then its components are continuous at x_0.

<u>Proof</u>: First, by the compactness of Θ and the continuity of $f(x,\theta)$ it follows that at least one vector function $\hat{\theta}(x) \in \Theta$ exists such that $f(x,\hat{\theta}(x))=\sup_{\theta\in\Theta} f(x,\theta)$. From lemma 2.3.1 we know that $f(x,\hat{\theta}(x))$ is continuous. Now suppose that

*) See also section 2.3.1 for related questions.

$\hat{\theta}(x)$ is unique. Let x_0 be a discontinuity point of $\hat{\theta}(x)$. Then there exists a sequence (x_j) of points in R^k such that

$x_j \to x_0$, $\hat{\theta}(x_j) \to \theta_1 \neq \hat{\theta}(x_0)$ for $j \to \infty$, while by the compactness of Θ it follows that $\theta_1 \in \Theta$. However, since $f(x,\theta)$ is continuous on $R^k \times \Theta$ and since $f(x,\hat{\theta}(x))$ is continuous on R^k we have:

$$f(x_j,\hat{\theta}(x_j)) \to f(x_0,\theta_1) \qquad \text{for } j \to \infty ,$$

$$f(x_j,\hat{\theta}(x_j)) \to f(x_0,\hat{\theta}(x_0)) \qquad \text{for } j \to \infty ,$$

hence $f(x_0,\theta_1)= f(x_0,\hat{\theta}(x_0))$. Since $\theta_1 \neq \hat{\theta}(x_0)$ and $f(x_0,\theta_1)=\sup_{\theta\in\Theta} f(x_0,\theta)$, it follows that $\hat{\theta}(x)$ is not unique at $x = x_0$, which contradicts the hypothesis. Thus if $\hat{\theta}(x)$ is unique then it is continuous. □

From lemma 3.1.1 it follows now that at least one of the solutions of (3.1.2) is a Borel measurable vector function of $(\underline{y}_1,\underline{x}_1),\ldots,(\underline{y}_n,\underline{x}_n)$, and hence a random vector.

3.1.2 Strong consistency

The strong consistency of $\hat{\underline{\theta}}_n$ will be proved by showing that there is a real function $Q(\theta)$ on Θ such that

$$\sup_{\theta\in\Theta} |\underline{Q}_n(\theta)-Q(\theta)|\to 0 \text{ a.s.} \qquad (3.1.4)$$

and that this limit function $Q(\theta)$ has a unique infimum at $\theta=\theta_0$, because then we can apply the following lemma:

> Lemma 3.1.3. Let $(\underline{Q}_n(\theta))$ be a sequence of random functions on a compact set Θ. Suppose $\underline{Q}_n(\theta)\to Q(\theta)$ a.s. pseudo-uniformly on Θ , where $Q(\theta)$ is a continuous non-random function on Θ . If there is a unique point θ_0 in Θ satisfying $Q(\theta_0)= \inf_{\theta\in\Theta} Q(\theta)$ then for any random vector $\hat{\underline{\theta}}_n$ a.s. in Θ satisfying $\underline{Q}_n(\hat{\underline{\theta}}_n)= \inf_{\theta\in\Theta}\underline{Q}_n(\theta)$, we have $\hat{\underline{\theta}}_n \to \theta_0$ a.s.

Proof: Let $\{\Omega,\mathfrak{F},P\}$ be the probability space. Let N the null set on which $\underline{Q}_n(\theta)\to Q(\theta)$ a.s. pseudo-uniformly on Θ fails to hold. Then for $\omega\in\Omega\setminus N$:

$$0 \le Q(\hat{\theta}_n(\omega)) - Q(\theta_0) = Q(\hat{\theta}_n(\omega)) - Q_n(\hat{\theta}_n(\omega),\omega) + Q_n(\hat{\theta}_n(\omega),\omega) - Q(\theta_0) \le$$

$$\le Q(\hat{\theta}_n(\omega)) - Q_n(\hat{\theta}_n(\omega),\omega) + Q_n(\theta_0,\omega) - Q(\theta_0) \le$$

$$\le 2 \sup_{\theta \in \text{Ⓗ}} |Q_n(\theta,\omega) - Q(\theta)| \to 0 \quad \text{as} \quad n \to \infty ,$$

hence for $\omega \in \Omega \setminus N$,

$$Q(\hat{\theta}_n(\omega)) \to Q(\theta_0) \text{ as } n \to \infty. \tag{3.1.5}$$

Let $\theta^*(\omega)$ be any limit point of the sequence $(\hat{\theta}_n(\omega))$. Since this sequence lies in the compact set Ⓗ, all its limit points lie in Ⓗ. Thus $\theta^*(\omega) \in$ Ⓗ . By definition of limit point there is a subsequence $(n_k(\omega))$ such that $\hat{\theta}_{n_k(\omega)}(\omega) \to \theta^*(\omega)$ as $k \to \infty$. Thus for every $\omega \in \Omega \setminus N$ we have:

$$Q(\hat{\theta}_{n_k(\omega)}(\omega)) \to Q(\theta^*(\omega)) = Q(\theta_0) \quad \text{as} \quad k \to \infty .$$

However, since θ_0 is unique we must have $\theta^*(\omega)=\theta_0$ and consequently since now all the limit points of $\hat{\theta}_n(\omega)$ are equal to θ_0 we have $\lim \hat{\theta}_n(\omega)=\theta_0$ for $\omega \in \Omega \setminus N$. This proves the lemma. □

<u>Remark</u>: Note that we have <u>not</u> used that $\underline{Q}_n(\hat{\underline{\theta}}_n)$ is a random variable for each n. Thus lemma 3.1.3 even holds if $\underline{Q}_n(\hat{\underline{\theta}}_n)$ is not a random variable. However, we shall only apply this lemma to cases where $\underline{Q}_n(\theta) \to Q(\theta)$ a.s. uniformly on Ⓗ and where $\underline{Q}_n(\hat{\underline{\theta}}_n)$ is a random variable.

For showing (3.1.4) we shall impose further conditions such that the second term in (3.1.3) converges a.s. to zero uniformly on Ⓗ and the third terms in (3.1.3) converges a.s. to the limit of its mathematical expectation, uniformly on Ⓗ . The first term converges a.s. to σ^2:

$$\frac{1}{n}\sum_{j=1}^{n}\underline{u}_j^2 \to \sigma^2 \quad \text{a.s.} , \tag{3.1.6}$$

because of Kolmogorov's strong law of large numbers (theorem 2.2.3B).
These further conditions are:

<u>Assumption 3.1.3</u>. The distribution functions $H_j(x)$, respectively, of the regressors satisfy:

$\frac{1}{n}\sum_{j=1}^{n} H_j(x) \rightarrow H(x)$ properly.

and :

<u>Assumption 3.1.4</u>.

a) $\sup_n \frac{1}{n}\sum_{j=1}^{n} E\{\overline{g(\underline{x}_j,\theta)}^{\textcircled{H}}\}^{2+\delta} < \infty$ for some $\delta > 0$
and
b) $E\{\overline{g(\underline{x}_j,\theta)}^{\textcircled{H}}\}^4 = O(j^{\mu})$ for some $\mu < 1$.

These assumptions will enable us to use theorem 2.3.3 for proving (3.1.4). First, observe from definition 2.3.2 and the inequalities (2.3.16)through (2.3.18) that :

$$\overline{\underline{u}_j(g(\underline{x}_j,\theta_0)-g(\underline{x}_j,\theta))}^{\textcircled{H}} \leq 2\,|\underline{u}_j|\{\overline{g(\underline{x}_j,\theta)}^{\textcircled{H}}\} \quad , \tag{3.1.7}$$

so that from the assumptions 3.1.1, 3.1.2 and 3.1.4 a:

$$\frac{1}{n}\sum_{j=1}^{n} E\{\overline{\underline{u}_j(g(\underline{x}_j,\theta_0)-g(\underline{x}_j,\theta))}^{\textcircled{H}}\}^{1+\delta^*} < \infty$$

for $\delta^* = 1+\delta$ and:

$$E\{\overline{\underline{u}_j(g(\underline{x}_j,\theta_0)-g(\underline{x}_j,\theta))}^{\textcircled{H}}\}^2 = O(j^{\mu^*})$$

for $\mu^* = \mu/2$. Moreover, assumption 3.1.3 implies:

$$\frac{1}{n}\sum_{j=1}^{n} F(u)H_j(x) \rightarrow F(u)H(x) \quad \text{properly.} \tag{3.1.8}$$

So it follows from theorem 2.3.3 that :

$$\frac{1}{n}\sum_{j=1}^{n} \underline{u}_j(g(\underline{x}_j,\theta_0)-g(\underline{x}_j,\theta)) \rightarrow \int u\,dF(u)\int(g(x,\theta_0)-g(x,\theta))dH(x) \text{ a.s. uniformly on } \textcircled{H}.$$

But from assumption 3.1.2 we have $\int u\,dF(u)=0$, hence:

$$\frac{1}{n}\sum_{j=1}^{n} \underline{u}_j(g(\underline{x}_j,\theta_0)-g(\underline{x}_j,\theta)) \rightarrow 0 \quad \text{a.s. uniformly on } \textcircled{H} \tag{3.1.9}$$

Next, it follows from the inequalities (2.3.16) through (2.3.18) that:

$$\overline{(g(\underline{x}_j,\theta_0)-g(\underline{x}_j,\theta))^2}^{\textcircled{H}} \leq 4.\overline{g(\underline{x}_j,\theta)^2}^{\textcircled{H}} \leq 4\{\overline{g(\underline{x}_j,\theta)}^{\textcircled{H}}\}^2 \quad \text{a.s.}, \tag{3.1.10}$$

so that from assumption 3.1.4a:

$$\sup_n \frac{1}{n}\sum_{j=1}^{n} E\{\overline{(g(\underline{x}_j,\theta_0)-g(\underline{x}_j,\theta))^2}^{\Theta}\}^{1+\delta} < \infty$$

and from assumption 3.1.4b:

$$E\{\overline{(g(\underline{x}_j,\theta_0)-g(\underline{x}_j,\theta))^2}^{\Theta}\}^2 = O(j^{\mu}).$$

So it follows from the assumptions 3.1.1 through 3.1.4 and from theorem 2.3.3 that

$$\frac{1}{n}\sum_{j=1}^{n}(g(\underline{x}_j,\theta_0)-g(\underline{x}_j,\theta))^2 \to \int (g(x,\theta_0)-g(x,\theta))^2 dH(x) \text{ a.s. uniformly on } \Theta \qquad (3.1.11)$$

Combining (3.1.6), (3.1.9) and (3.1.11) we now see that under the assumptions 3.1.1 through 3.1.4, (3.1.4) is satisfied, where

$$Q(\theta) = \sigma^2 + \int (g(x,\theta_0)-g(x,\theta))^2 dH(x) . \qquad (3.1.12)$$

It is not hard to see that the limit function $Q(\theta)$ is continuous on Θ. Moreover, if

<u>Assumption 3.1.5.</u> $\int (g(x,\theta_0)-g(x,\theta))^2 dH(x) = 0$ if and only if $\theta = \theta_0$,

then θ_0 is a unique point in Θ satisfying:

$$Q(\theta_0) = \inf_{\theta\in\Theta} Q(\theta). \qquad (3.1.13)$$

Finally, since $\underline{Q}_n(\theta)$ is a continuous function of the random variables $\underline{u}_1,\dots,\underline{u}_n,\underline{x}_1,\dots\underline{x}_n$ and the parameter θ, $\underline{Q}_n(\hat{\underline{\theta}}_n)$ is a random variable.

Applying lemma 3.1.3, we now conclude:

<u>Theorem 3.1.1.</u> Under the assumptions 3.1.1 through 3.1.5 the least squares estimator $\hat{\underline{\theta}}_n$ is strongly consistent: $\hat{\underline{\theta}}_n \to \theta_0$ a.s.

<u>Remark 1</u>: The assumption that the regressors are random vectors or variables is not essential, for nonrandom vectors x_j can be considered as independent random vectors $\underline{x}_j$ with distribution function $H_j(x) = I(x_j \le x)$, where I is the indicator function of the set $\{x_j \le x\}$:

$$I\{x_j \le x\} = \begin{cases} 1 \text{ if } x_j \le x \\ 0 \text{ if not.} \end{cases}$$

Then assumption 3.1.3 requires that the empirical distribution function of $\{x_1,\dots,x_n\}$ converges properly.

Remark 2: There is a close connection between the assumptions 3.1.3, 3.1.4 and 3.1.5 and the usual assumptions which have to be made for establishing consistency of the linear least squares estimator.

Let $g(x,\theta)= x'\theta$,(thus p=q), $\underline{X}=\begin{pmatrix}\underline{x}_1'\\ \vdots \\ \underline{x}_n'\end{pmatrix}$. Then from part b) of assumption 3.1.4 and the strong law of large numbers (theorem 2.2.3A)

$$\frac{1}{n}\underline{X}'\underline{X} - E\frac{1}{n}\underline{X}'\underline{X} \to O \quad \text{a.s.} ,$$

where O is a zero matrix, and from assumption 3.1.3, part a) of assumption 3.1.4. and theorem 2.2.15,

$$E\frac{1}{n}\underline{X}'\underline{X} \quad \to \quad R_* \text{ , where } R_*=(r_{*,i_1,i_2})=\left(\int x^*_{i_1} x^*_{i_2}\, dH(x^*)\right),$$

with $x^*_{i_1}$ and $x^*_{i_2}$ components of the vector x^*.

Thus:

$$\frac{1}{n}\underline{X}'\underline{X} \to R_* \quad \text{a.s.}$$

Moreover, assumption 3.1.5 says that R_* is positive definite.

3.1.3 Asymptotic normality

The limiting distribution of the least squares estimator can be derived in a similar way as for the maximum likelihood estimator, namely using the Taylor expansion of the vector of first derivatives of the objective function:

$$(\partial/\partial\theta)\underline{Q}_n(\hat{\underline{\theta}}_n)= (\partial/\partial\theta)\underline{Q}_n(\theta_0)+(\hat{\underline{\theta}}_n- \theta_0)'(\partial/\partial\theta)(\partial/\partial\theta')\underline{Q}_n(\underline{\theta}^*_n)\ ^{*)} , \tag{3.1.14}$$

where $\underline{\theta}^*_n$ is some mean value satisfying

$$|\underline{\theta}^*_n - \theta_0| \leq |\hat{\underline{\theta}}_n - \theta_0| . \tag{3.1.15}$$

Thus we first show that plim $\sqrt{n}(\partial/\partial\theta)\underline{Q}_n(\hat{\underline{\theta}}_n)=0'$, where $0'$ is a zero row vector; then we show that $(\partial/\partial\theta)(\partial/\partial\theta')\underline{Q}_n(\underline{\theta}^*_n)$ converges a.s. to a positive definite matrix and finally we show that $\sqrt{n}(\partial/\partial\theta')\underline{Q}_n(\theta_0)$ converges in distribution to the normal with zero mean vector.

---

*) By $(\partial/\partial\theta)f(\theta)$ we denote the row vector of partial derivative of $f(\theta)$ and by $(\partial/\partial\theta')f(\theta)$ the column vector of partial derivatives. Moreover, $(\partial/\partial\theta)(\partial/\partial\theta')f(\theta)$ denotes the matrix $((\partial/\partial\theta_{i_1})(\partial/\partial\theta_{i_2})f(\theta))$.

Before doing this we first question whether or not $\underline{\theta}_n^*$ is a random vector, because only if so then (3.1.15) and theorem 3.1.1 imply

$\underline{\theta}_n^* \to \theta_0$ a.s.

The answer is yes: $\underline{\theta}_n^*$ is indeed a random vector, as follows from lemma 3 of Jennrich (1969):

> Lemma 3.1.4. Let $f(x,\theta)$ be a real valued function on $R^k \times \Theta$, where Θ is a convex compact subset of R^m. For each θ in Θ let $f(x,\theta)$ be a Borel measurable function on R^k and for each $x \in R^k$ let $f(x,\theta)$ be a continuous differentiable function on Θ. Let $\theta_1(x)$ and $\theta_2(x)$ be Borel measurable functions from R^k into Θ. Then there exists a Borel measurable function $\theta^*(x)$ from R^k into Θ such that
>
> (i) $f(x,\theta_1(x)) - f(x,\theta_2(x)) = (\partial/\partial\theta) f(x,\theta^*(x))(\theta_1(x) - \theta_2(x))$,
>
> (ii) $\theta^*(x)$ lies on the segment joining $\theta_1(x)$ and $\theta_2(x)$.

Proof: Jennrich (1969).

Remark: Conclusion (ii) means that there is a (Borel measurable) real function $\lambda(x)$ from R^k into $[0,1]$ such that

$$\theta^*(x) = \theta_1(x) + \lambda(x)[\theta_2(x) - \theta_1(x)] \qquad (3.1.17)$$

Thus if Θ is convex compact then $\underline{\theta}_n^*$ is a random vector because $\underline{\theta}_n^*$ is a Borel measurable vector function of the observations.

Now we show plim $\sqrt{n}(\partial/\partial\theta)\,\underline{Q}_n(\hat{\underline{\theta}}_n) = 0'$. Let θ_0 be an interior point of Θ. Then by theorem 3.1.1 the least squares estimator $\hat{\underline{\theta}}_n$ is a.s. an interior point of Θ, hence $(\partial/\partial\theta)\underline{Q}_n(\hat{\underline{\theta}}_n) = 0'$ a.s. for large n, where 0' is a zero row vector, and consequently $b_n(\partial/\partial\theta)\underline{Q}_n(\hat{\underline{\theta}}_n) \to 0'$ a.s. for every sequence (b_n). Thus $\sqrt{n}(\partial/\partial\theta)\underline{Q}_n(\hat{\underline{\theta}}_n) \to 0'$ a.s. and hence also in probability.

For later use we shall formulate this result more generally as follows:

> Lemma 3.1.5. Let $(\underline{Q}_n(\theta))$ be a sequence of continuous differentiable random functions on a compact set Θ. Let the sequence $(\hat{\underline{\theta}}_n)$ of random vectors in Θ be obtained from $\underline{Q}_n(\hat{\underline{\theta}}_n) = \inf_{\theta\in\Theta} \underline{Q}_n(\theta_n)$. If $\hat{\underline{\theta}}_n$

converges a.s. to an interior point of Ⓗ then for every sequence(b_n) we have $b_n(\partial/\partial\theta)\underline{Q}_n(\hat{\underline{\theta}}_n) \to 0'$ a.s., provided that $(\partial/\partial\theta)\underline{Q}_n(\hat{\underline{\theta}}_n)$ is a random vector for each n.

Proof: We give now an exact proof. Let$\{\Omega,\mathfrak{F},P\}$ be the probability space. Then for every $\omega\in\Omega\setminus N$, where N is some null set, we can find a number $n_0(\omega)$ such that $\hat{\theta}_n(\omega)$ is an interior point of Ⓗ if $n\geq n_0(\omega)$. Since $Q_n(\hat{\theta}_n(\omega),\omega)=\inf_{\theta\in Ⓗ} Q_n(\theta,\omega)$ it follows now that $(\partial/\partial\theta)Q_n(\hat{\theta}_n(\omega),\omega)=0'$ if $n\geq n_0(\omega)$ and hence that for every sequence b_n, $b_n(\partial/\partial\theta)Q_n(\hat{\theta}_n(\omega),\omega)=0'$ if $n\geq n_0(\omega)$, which proves the lemma. □

So we have that if

Assumption 3.1.6. Ⓗ is convex and θ_0 is an interior point of it,

then under the conditions of theorem 3.1.1

$$\sqrt{n}(\partial/\partial\theta)\underline{Q}_n(\theta_0)+\sqrt{n}(\hat{\underline{\theta}}_n-\theta_0)'(\partial/\partial\theta)(\partial/\partial\theta')\underline{Q}_n(\underline{\theta}_n^*) \to 0' \quad \text{a.s.} \tag{3.1.18}$$

The next step in deriving the limiting distribution of the least squares estimator is to show that $(\partial/\partial\theta)(\partial/\partial\theta')\underline{Q}_n(\underline{\theta}_n^*)$ converges in pr. to a non-random matrix. Consider an element of the matrix of partial derivatives of $\underline{Q}_n(\theta)$:

$$(\partial/\partial\theta_{i_1})(\partial/\partial\theta_{i_2})\underline{Q}_n(\theta)=2\frac{1}{n}\sum_{j=1}^n\{(\partial/\partial\theta_{i_1})g(\underline{x}_j,\theta)\}\{(\partial/\partial\theta_{i_2})g(\underline{x}_j,\theta)\}$$

$$-2\frac{1}{n}\sum_{j=1}^n(\underline{u}_j+g(\underline{x}_j,\theta_0)-g(\underline{x}_j,\theta))(\partial/\partial\theta_{i_1})(\partial/\partial\theta_{i_2})g(\underline{x}_j,\theta)$$

$$=2\underline{a}_{n,i_1,i_2}(\theta)-2\underline{b}_{n,i_1,i_2}(\theta)\quad, \tag{3.1.19}$$

say. If:

Assumption 3.1.7.

a) $\sup_n \frac{1}{n}\sum_{j=1}^n E\{\overline{(\partial/\partial\theta_i)g(\underline{x}_j,\theta)}^{Ⓗ}\}^{2+\delta}<\infty$

and

b) $E\{\overline{(\partial/\partial\theta_i)g(\underline{x}_j,\theta)}^{Ⓗ}\}^4=O(j^\mu)$

for $i=1,2,\ldots,q$ and some $\delta>0,\mu<1$, respectively, then from the inequalities (2.3.16), (2.3.17) and the trivial inequality $|ab| \leq \frac{1}{2}(a^2+b^2)$,

$$\sup_n \frac{1}{n}\sum_{j=1}^n E(\overline{\{(\partial/\partial\theta_{i_1})g(\underline{x}_j,\theta)\}\{(\partial/\partial\theta_{i_2})g(\underline{x}_j,\theta)\}}^{\Theta})^{1+\delta^*} < \infty,$$

$$E(\overline{\{(\partial/\partial\theta_{i_1})g(\underline{x}_j,\theta)\}\{\partial/\partial\theta_{i_2})g(\underline{x}_j,\theta)\}}^{\Theta})^2 = O(j^{\mu}) \tag{3.1.20}$$

for $\delta^*=\delta/2$, so that by assumption 3.1.1 (the independence of the $\underline{x}_j$'s), assumption 3.1.3 and theorem 2.3.3,

$$\sup_{\theta\in\Theta} |\underline{a}_{n,i_1,i_2}(\theta) - a_{i_1,i_2}(\theta)| \to 0 \text{ a.s.}, \tag{3.1.21}$$

where

$$a_{i_1,i_2}(\theta) = \int\{(\partial/\partial\theta_{i_1})g(x,\theta)\}\{(\partial/\partial\theta_{i_2})g(x,\theta)\}\,dH(x). \tag{3.1.22}$$

Consequently we have by (3.1.16)

$$\underline{a}_{n,i_1,i_2}(\underline{\theta}_n^*) \to a_{i_1,i_2}(\theta_0) \text{ a.s.}, \tag{3.1.23}$$

because of theorem 2.3.2 and the continuity of $a_{i_1,i_2}(\theta)$ on Θ.

Moreover, if

<u>Assumption 3.1.8</u>. $\sup_n \frac{1}{n}\sum_{j=1}^n E(\overline{(\partial/\partial\theta_{i_1})(\partial/\partial\theta_{i_2})g(\underline{x}_j,\theta)}^{\Theta})^{2+\delta} < \infty$

for $i_1,i_2 = 1,2,\ldots,q$ and some $\delta > 0$,

then from assumption 3.1.4a, the inequalities (2.3.16) and (2.3.17) and from Liapounov's inequality

$$\sup_n \frac{1}{n}\sum_{j=1}^n E\{|\underline{u}_j|(\overline{(\partial/\partial\theta_{i_1})(\partial/\partial\theta_{i_2})g(\underline{x}_j,\theta)}^{\Theta})\}^{1+\delta} < \infty,$$

$$\sup_n \frac{1}{n}\sum_{j=1}^n E(\overline{(g(\underline{x}_j,\theta_0) - g(\underline{x}_j,\theta))(\partial/\partial\theta_{i_1})(\partial/\partial\theta_{i_2})g(\underline{x}_j,\theta)}^{\Theta})^{1+\delta} < \infty, \tag{3.1.24}$$

for some $\delta > 0$, so that from theorem 2.3.4

$$\text{plim} \sup_{\theta\in\Theta} |\underline{b}_{n,i_1,i_2}(\theta) - b_{i_1,i_2}(\theta)| = 0, \tag{3.1.25}$$

where $b_{n,i_1,i_2}(\theta)$ is defined in (3.1.19) and

$$b_{i_1,i_2}(\theta)=\int (u+g(x,\theta_0)-g(x,\theta))(\partial/\partial\theta_{i_1})(\partial/\partial\theta_{i_2})g(x,\theta)d\{F(u)H(x)\}$$

$$=\int (g(x,\theta_0)-g(x,\theta))(\partial/\partial\theta_{i_1})(\partial/\partial\theta_{i_2})g(x,\theta)dH(x) \qquad (3.1.26)$$

(the simplification involved follows from $\int udF(u)=0$). Since obviously $b_{i_1,i_2}(\theta_0)=0$, it follows from (3.1.16) and (3.1.25) that

$$\underline{b}_{n,i_1,i_2}(\underline{\theta}_n^*) \to 0 \text{ in pr. .} \qquad (3.1.27)$$

Summarizing we thus have:

Lemma 3.1.6. Under the assumptions of theorem 3.1.1 and the additional assumptions 3.1.6, 3.1.7 and 3.1.8 we have
$(\partial/\partial\theta)(\partial/\partial\theta')\underline{Q}_n(\underline{\theta}_n^*) \to 2A(\theta_0)$ in pr. ,
where $A(\theta)=(a_{i_1,i_2}(\theta))=(\int\{(\partial/\partial\theta_{i_1})g(x,\theta)\}\{(\partial/\partial\theta_{i_2})g(x,\theta)\}dH(x))$.

Lemma 3.1.7. Under the assumptions 3.1.1, 3.1.3 and 3.1.7
$\underline{A}_n(\theta)\to A(\theta)$ a.s. uniformly on Ⓗ , where
$\underline{A}_n(\theta)=(\underline{a}_{n,i_1,i_2}(\theta))=(\frac{1}{n}\sum_{j=1}^n\{(\partial/\partial\theta_{i_1})g(\underline{x}_j,\theta)\}\{(\partial/\partial\theta_{i_2})g(\underline{x}_j,\theta)\})$.

Now we show the asymptotic normality of
$\sqrt{n}(\partial/\partial\theta)\underline{Q}_n(\theta_0)=\frac{1}{\sqrt{n}}\sum_{j=1}^n -2\underline{u}_j(\partial/\partial\theta)g(\underline{x}_j,\theta_0)$.

Since $\underline{u}_j$ and $\underline{x}_j$ are mutually independent it follows from assumption 3.1.2 that

$$E\sqrt{n}(\partial/\partial\theta')\underline{Q}_n(\theta_0)=0' \qquad (3.1.28)$$

and since the sequence of pairs $(\underline{u}_j,\underline{x}_j)$ is independent we have:

$$\text{var}(\sqrt{n}(\partial/\partial\theta')\underline{Q}_n(\theta_0))=4\sigma^2\frac{1}{n}\sum_{j=1}^n E\{(\partial/\partial\theta')g(\underline{x}_j,\theta_0)\}\{(\partial/\partial\theta)g(\underline{x}_j,\theta_0)\}$$

$$=4\sigma^2 E\,\underline{A}_n(\theta_0)\to 4\sigma^2 A(\theta_0) , \qquad (3.1.29)$$

where the latter result follows from a similar argument as is used for proving lemma 3.1.7.

Let ζ be any vector in R^q and put

$$\underline{w}_j = -2\underline{u}_j(\partial/\partial\theta)g(\underline{x}_j,\theta_0)\zeta ,$$

$$\underline{s}_n = \sum_{j=1}^n \underline{w}_j .$$

Then the $\underline{w}_j$'s are independent random variables with zero expectations and

$\lim \frac{1}{n}\mathrm{var}(\underline{s}_n) = 4\sigma^2\zeta' A(\theta_0)\zeta$. Suppose that:

<u>Assumption 3.1.9</u>. $\det(A(\theta_0)) \neq 0$.

Since $A(\theta_0)$ is the limit of a sequence of positive semidefinite matrices and hence itself positive semidefinite, this assumption implies that $A(\theta_0)$ is positive definite: $\zeta' A(\theta_0)\zeta > 0$ for every nonzero vector $\zeta \in R^q$.

So, using (2.1.13) and the assumption 3.1.2 and 3.1.7, it follows that

$$\sum_{j=1}^n E\left|\frac{\underline{w}_j}{\sqrt{\mathrm{var}(\underline{s}_n)}}\right|^{2+\delta} = \frac{2^{2+\delta}\int |u|^{2+\delta}dF(u)\frac{1}{n}\sum_{j=1}^n E|(\partial/\partial\theta)g(\underline{x}_j,\theta_0)\zeta|^{2+\delta}}{n^{\frac{\delta}{2}}(\frac{1}{n}\mathrm{var}(\underline{s}_n))^{\frac{2+\delta}{2}}}$$

$$\leq \frac{2^{2+\delta}|\zeta|^{2+\delta}\int |u|^{2+\delta}dF(u)q^{\delta}\sum_{i=1}^q \sup_n \frac{1}{n}\sum_{j=1}^n E((\partial/\partial\theta_i)g(\underline{x}_j,\theta_0))^{1+\frac{\delta}{2}}}{n^{\frac{\delta}{2}}(\frac{1}{n}\mathrm{var}(s_n))^{\frac{2+\delta}{2}}}$$

$$= O(n^{-\frac{\delta}{2}}) \to 0 , \tag{3.1.30}$$

so that by theorem 2.4.6

$$\sqrt{n}(\partial/\partial\theta)\underline{Q}_n(\theta_0)\zeta = \frac{1}{\sqrt{n}}\sum_{j=1}^n \underline{w}_j \to N(0,4\sigma^2\zeta' A(\theta_0)\zeta) \text{ in distr. for every } \zeta \in R^q.$$

Consequently by theorem 2.4.2,

$$\sqrt{n}(\partial/\partial\theta')\underline{Q}_n(\theta_0) \to N_q(0,4\sigma^2 A(\theta_0)) \text{ in distr. .} \tag{3.1.31}$$

Since $A(\theta_0)$ is non-singular and the elements of the inverse of a matrix are continuous functions of the matrix involved it follows from (3.1.18), lemma 3.1.6 and theorem 2.2.5 that:

$$\sqrt{n}(\underline{\hat{\theta}}_n - \theta_0) - [(\partial/\partial\theta)(\partial/\partial\theta')\underline{Q}_n(\underline{\theta}_n^*)]^{-1}\sqrt{n}(\partial/\partial\theta')\underline{Q}_n(\theta_0) \to 0 \text{ in pr.} \quad (3.1.32)$$

and from (3.1.31), lemma 3.1.6 and theorem 2.2.14 that

$$[(\partial/\partial\theta)(\partial/\partial\theta')Q_n(\underline{\theta}_n^*)]^{-1}\sqrt{n}(\partial/\partial\theta')\underline{Q}_n(\theta_0) \to N_q(0,\sigma^2 A(\theta_0)^{-1}) \text{ in distr.}. \quad (3.1.33)$$

From (3.1.32), (3.1.33) and theorem 2.2.11 it follows now that:

$$\sqrt{n}(\underline{\hat{\theta}}_n - \theta_0) \to N_q(0,\sigma^2 A(\theta_0)^{-1}) \text{ in distr.}. \quad (3.1.34)$$

Note that by lemma 3.1.6 and the nonsingularity of $A(\theta_0)$ it follows that under the conditions of theorem 3.1.1, $\underline{A}_n(\underline{\hat{\theta}}_n)^{-1} \to A(\theta_0)^{-1}$ a.s. while from (3.1.4) and the theorems 3.1.1 and 2.3.3 we have
$\underline{Q}_n(\underline{\hat{\theta}}_n) \to Q(\theta_0) = \sigma^2$ a.s. Hence:

$$\underline{Q}_n(\underline{\hat{\theta}}_n)\underline{A}_n(\underline{\hat{\theta}}_n)^{-1} \to \sigma^2 A(\theta_0)^{-1} \text{ a.s.}. \quad (3.1.35)$$

We have proved by now:

<u>Theorem 3.1.2</u>. Under the assumptions of theorem 3.1.1 and the additional assumptions 3.1.6 through 3.1.9 we have

$$\sqrt{n}(\underline{\hat{\theta}}_n - \theta_0) \to N_q(0, \sigma^2 A(\theta_0)^{-1}) \text{ in distr.}$$

and

$$\underline{Q}_n(\underline{\hat{\theta}}_n)\underline{A}_n(\underline{\hat{\theta}}_n)^{-1} \to \sigma^2 A(\theta_0)^{-1} \text{ a.s.}$$

3.1.4 Weak consistency and asymptotic normality under weaker conditions

The almost sure results in the theorems 3.1.1 and 3.1.2 are obtained by applying theorem 2.3.3 to the case involved. But if we only assume part a) of assumption 3.1.4 then by applying theorem 2.3.4, (3.1.9) now becomes

$$\text{plim} \sup_{\theta \in Ⓗ} \left|\frac{1}{n}\sum_{j=1}^n \underline{u}_j(g(\underline{x}_j,\theta_0) - g(\underline{x}_j,\theta))\right| = 0 \quad (3.1.36)$$

and (3.1.11) becomes:

$$\text{plim} \sup_{\theta \in Ⓗ} \left|\frac{1}{n}\sum_{j=1}^n (g(\underline{x}_j,\theta_0) - g(\underline{x}_j,\theta))^2 - \int (g(x,\theta_0) - g(x,\theta))^2 dH(x)\right| = 0 . \quad (3.1.37)$$

Since (3.1.6) still holds because of the assumptions 3.1.1 and 3.1.2, it follows from (3.1.6), (3.1.36) and (3.1.37) that (3.1.4) now becomes:

$$\text{plim} \sup_{\theta \in \Theta} |\underline{Q}_n(\theta) - Q(\theta)| = 0 . \tag{3.1.38}$$

Moreover, combining theorem 2.2.4 and lemma 3.1.3 it is easily proved that

<u>Lemma 3.1.8</u>. If the conditions of lemma 3.1.3 are satisfied, except that now $\underline{Q}_n(\theta) \to Q(\theta)$ in pr. pseudo-uniformly on Θ, then plim $\hat{\underline{\theta}}_n = \theta_0$.

Thus we have:

<u>Theorem 3.1.3</u>. Under the assumptions 3.1.1, 3.1.2, 3.1.3, 3.1.4a and 3.1.5 the least squares estimator $\hat{\underline{\theta}}_n$ is weakly consistent: plim $\hat{\underline{\theta}}_n = \theta_0$.

Similarly, only assuming part a) of assumption 3.1.7 it follows from theorem 2.3.4 that (3.1.21) now becomes

$$\text{plim} \sup_{\theta \in \Theta} |\underline{a}_{n,i_1,i_2}(\theta) - a_{i_1,i_2}(\theta)| = 0 \tag{3.1.39}$$

Since (3.1.25), (3.1.28) and (3.1.30) remain valid, while under the assumptions of theorem 3.1.3 it follows from (3.1.15) that now plim $\underline{\theta}_n^* = \theta_0$ instead of (3.1.16), we thus conclude (again with theorem 2.2.4 in mind):

<u>Theorem 3.1.4</u>. Under the assumptions of theorem 3.1.3 and the assumptions 3.1.6, 3.1.7a, 3.1.8 and 3.1.9, $\sqrt{n}(\hat{\underline{\theta}}_n - \theta_0) \to N_q(0, \sigma^2 A(\theta_0)^{-1})$ in distr. and plim $\underline{Q}_n(\hat{\underline{\theta}}_n)\underline{A}_n(\hat{\underline{\theta}}_n)^{-1} = \sigma^2 A(\theta_0)^{-1}$.

3.1.5 <u>Asymptotic properties if the error distribution has infinite variance. Symmetric stable error distributions</u>

In this section it will be shown that the least squares estimator remains weakly consistent if the assumption $\int |u|^{2+\delta} dF(u) < \infty$ for some $\delta > 0$ (see assumption 3.1.2) is weakened to

$$\int |u|^{1+\delta} dF(u) < \infty \text{ for some } \delta > 0.$$

This applies to symmetric stable error distributions with characteristic exponent greater than 1. For that case also the limiting distribution of the least squares estimator will be derived, which turns out to be nonnormal.

For proving weak consistency under the weaker condition above, observe that the least squares estimator $\hat{\underline{\theta}}_n$ can also be obtained from

$$\underline{Q}_n^0(\hat{\underline{\theta}}_n)= \inf_{\theta\in Ⓗ} \underline{Q}_n^0(\theta) \quad , \quad \theta\in Ⓗ \ , \tag{3.1.40}$$

where :

$$\underline{Q}_n^0(\theta)= \underline{Q}_n(\theta)- \frac{1}{n} \textstyle\sum_{j=1}^n \underline{u}_j^2$$

$$= 2\frac{1}{n}\textstyle\sum_{j=1}^n \underline{u}_j(g(\underline{x}_j,\theta_0)- g(\underline{x}_j,\theta))+ \frac{1}{n} \textstyle\sum_{j=1}^n(g(\underline{x}_j,\theta_0)- g(\underline{x}_j,\theta))^2 \ . \tag{3.1.41}$$

Moreover, it is not hard to see that in this case (3.1.36) and (3.1.37) carry over, so that

$$\text{plim} \sup_{\theta\in Ⓗ} |\underline{Q}_n^0(\theta)- Q^0(\theta)| = 0 \ , \tag{3.1.42}$$

where

$$Q^0(\theta)= \int(g(x,\theta_0)- g(x,\theta))^2dH(x) \ , \tag{3.1.43}$$

because for proving (3.1.36) and (3.1.37) we did need the following parts of assumption 3.1.2: $\int udF(u)=0$ and $\int|u|^{1+\delta}dF(u) < \infty$ for some $\delta > 0$. Thus similar to theorem 3.1.3 we have

Theorem 3.1.5. Let the assumption of theorem 3.1.3 be satisfied except assumption 3.1.2 which is weakened as follows:

$$\int udF(u)=0,\int|u|^{1+\delta}dF(u)< \infty \text{ for some } \delta > 0. \tag{3.1.44}$$

Then still plim $\hat{\underline{\theta}}_n = \theta_0$.

Consequently if for example the error distribution is symmetric stable with characteristic exponent $\alpha\in(1,2)$ then the least squares estimator is still weakly consistent because in that case (3.1.44) holds for $0 < \delta < \alpha - 1$.*) But it is also possible to derive the limiting distribution in that case. First observe that lemma 3.1.6 carries over. Next put again

$\underline{w}_j= -2\underline{u}_j(\partial/\partial\theta)g(\underline{x}_j,\theta_0)\zeta$, $\underline{s}_n=\sum_{j=1}^n \underline{w}_j= n(\partial/\partial\theta)\underline{Q}_n(\theta_0)\zeta$, where $\zeta \in R^q$. Since the characteristic function of a symmetric stable distribution F(u) with characteristic exponent α is of the type

*) See section 2.4.

$$\int e^{itu}\, dF(u) = e^{-\sigma|t|^{\alpha}}, \quad \sigma > 0,\ \alpha \in (0,2] \tag{3.1.45}$$

and since the $\underline{u}_j$'s and the $\underline{x}_j$'s are independent it follows that:

$$Ee^{it\underline{s}_n} = \prod_{j=1}^{n} Ee^{it\underline{w}_j} = \prod_{j=1}^{n} Ee^{-\sigma|t|^{\alpha}\, 2^{\alpha}|(\partial/\partial\theta)g(\underline{x}_j,\theta_0)\zeta|^{\alpha}}$$

$$= Ee^{-\sigma|t|^{\alpha}\, 2^{\alpha} \sum_{j=1}^{n} |(\partial/\partial\theta)g(\underline{x}_j,\theta_0)\zeta|^{\alpha}}.$$

Hence, putting $t = n^{-\frac{1}{\alpha}}$,

$$Ee^{i.n^{1-\frac{1}{\alpha}} (\partial/\partial\theta)\underline{Q}_n(\theta_0)\zeta} = Ee^{-\sigma 2^{\alpha} \frac{1}{n} \sum_{j=1}^{n}|(\partial/\partial\theta)g(\underline{x}_j,\theta_0)\zeta|^{\alpha}}. \tag{3.1.46}$$

From theorem 2.3.4 and the assumptions 3.1.3 and 3.1.7a it follows that:

$$\operatorname{plim} \frac{1}{n} \sum_{j=1}^{n}|(\partial/\partial\theta)g(\underline{x}_j,\theta_0)\zeta|^{\alpha} = \int |(\partial/\partial\theta)g(x,\theta_0)\zeta|^{\alpha} dH(x),$$

and thus also

$$\operatorname{plim} e^{-\sigma 2^{\alpha} \frac{1}{n} \sum_{j=1}^{n}|(\partial/\partial\theta)g(\underline{x}_j,\theta_0)\zeta|^{\alpha}} = e^{-\sigma 2^{\alpha} \int |(\partial/\partial\theta)g(x,\theta_0)\zeta|^{\alpha} dH(x)}. \tag{3.1.47}$$

But since the left side random variable is uniformly bounded it follows from the bounded convergence theorem that the limit of (3.1.46) equals (3.1.47):

$$\lim Ee^{in^{1-\frac{1}{\alpha}} (\partial/\partial\theta)\underline{Q}_n(\theta_0)\zeta} = e^{-\sigma 2^{\alpha} \int |(\partial/\partial\theta_0)g(x,\theta_0)\zeta|^{\alpha} dH(x)}. \tag{3.1.48}$$

It follows now from (3.1.48), (3.1.14), theorem 2.2.14 and the fact that also in this case

$$n^{1-\frac{1}{\alpha}} (\partial/\partial\theta)\underline{Q}_n(\hat{\underline{\theta}}_n) \to 0' \text{ in pr.}; (\partial/\partial\theta)(\partial/\partial\theta')\underline{Q}_n(\underline{\theta}_n^{*}) \to 2A(\theta_0) \text{ in pr.}, \tag{3.1.49}$$

that

$$\lim Ee^{i.n^{1-\frac{1}{\alpha}} (\hat{\underline{\theta}}_n - \theta_0)'\zeta} = e^{-\sigma \int |(\partial/\partial\theta)g(x,\theta_0)A(\theta_0)^{-1}\zeta|^{\alpha} dH(x)}. \tag{3.1.50}$$

Summarizing we now have:

Theorem 3.1.6. Let the assumptions of the theorems 3.1.3 and 3.1.4 be satisfied, except assumption 3.1.2. If in addition the error distribution is symmetric stable with characteristic exponent $\alpha\in(1,2)$ and scaling parameter $\sigma > 0$, then $n^{1-\frac{1}{\alpha}}(\hat{\underline{\theta}}_n - \theta_0)$ converges in distribution to a limit-distribution with characteristic function as in (3.1.50).

However, since now $n^{1-\frac{1}{\alpha}} < \sqrt{n}$, the rate of convergence is lower than in the case of theorem 3.1.2, so that the least squares estimation method is unattractive when the error distribution is symmetric stable with characteristic exponent $\alpha < 2$.

Remark 3: If the conditions of theorem 3.1.5 are satisfied but $\int u^2 dF(u) = \infty$, then it is not hard to prove that $\text{plim}\ \frac{1}{n}\sum_{j=1}^n(\underline{y}_j - g(\underline{x}_j, \hat{\underline{\theta}}_n))^2 = \text{plim}\ \frac{1}{n}\sum_{j=1}^n \underline{u}_j^2 = +\infty$ and that the coëfficient of determination

$$\underline{R}_n^2 = 1 - \frac{\frac{1}{n}\sum_{j=1}^n(\underline{y}_j - g(\underline{x}_j, \hat{\underline{\theta}}_n))^2}{\frac{1}{n}\sum_{j=1}^n \underline{y}_j^2 - (\frac{1}{n}\sum_{j=1}^n \underline{y}_j)^2}$$

satisfies

$$\text{plim}\ \underline{R}_n^2 = 0 \ .$$

So if the standard error of our regression is high and the coefficient of determination is low, this may be caused by infinite variance of the error distribution.

If the error distribution is symmetric stable with characteristic exponent $\alpha\in(0,1]$, then the least squares estimator is not generally consistent. For example consider the simple linear regression model

$$\underline{y}_j = \theta_0 x_j + \underline{u}_j \ , j=1,2,\ldots,n, \ldots,$$

where the x_j's are non-random scalar regressors (hence θ_0 is a scalar too) and the $\underline{u}_j$'s are random drawings from a symmetric stable distribution with scaling parameter $\sigma > 0$ and characteristic exponent $\alpha\in(0,1]$.

Then $\hat{\underline{\theta}}_n - \theta_0 = \dfrac{\sum_{j=1}^n \underline{u}_j x_j}{\sum_{j=1}^n x_j^2}$, hence

$$Ee^{it(\hat{\theta}_n-\theta_0)} = \prod_{j=1}^{n} e^{-\sigma|t|^{\alpha}|x_j|^{\alpha}/|\sum_{j=1}^{n}x_j^2|^{\alpha}}$$

$$= e^{-\sigma|t|^{\alpha}n^{1-\alpha}\frac{1}{n}\sum_{j=1}^{n}|x_j|^{\alpha}/|\frac{1}{n}\sum_{j=1}^{n}x_j^2|^{\alpha}} .$$

Since $\hat{\theta}_n \to \theta_0$ in pr. is equivalent with $\hat{\theta}_n - \theta_0 \to 0$ in distr. (see the theorems 2.2.11 and 2.2.12) a necessary and sufficient condition for the consistency of $\hat{\theta}_n$ is

$$1 = \lim_{n\to\infty} Ee^{it(\hat{\theta}_n-\theta_0)} = e^{-\sigma|t|^{\alpha}\lim[n^{1-\alpha}\frac{1}{n}\sum_{j=1}^{n}|x_j|^{\alpha}/|\frac{1}{n}\sum_{j=1}^{n}x_j^2|^{\alpha}]}$$

for every t; hence

$$\lim[n^{1-\alpha}\frac{1}{n}\sum_{j=1}^{n}|x_j|^{\alpha}/|\frac{1}{n}\sum_{j=1}^{n}x_j^2|^{\alpha}]=0 . \tag{3.1.51}$$

But if in this case the assumptions 3.1.3 [with $H_j(x)$ the indicator function $I(x_j \leq x)$] and 3.1.4a hold, then by theorem 2.2.15,

$$\lim \frac{1}{n}\sum_{j=1}^{n}|x_j|^{\alpha} = \int |x|^{\alpha}dH(x), \quad \lim \frac{1}{n}\sum_{j=1}^{n}x_j^2 = \int x^2 dH(x),$$

so that (3.1.51) does not hold if these integrals are non-zero and $\alpha \leq 1$.

3.2 A class of nonlinear robust M-estimators

3.2.1 Introduction

We have just seen that the least squares estimation method is unattractive when the error distribution is symmetric stable with characteristic exponent $\alpha\in(1,2)$ and often inconsistent when $\alpha\in(0,1]$. In this case the maximum likelihood method should perform much better, but in practice it cannot be carried out because only for the cases $\alpha=1$ and $\alpha=2$ explicit forms of the densities of stable distributions are known. However, a way out of the problem is provided by the so-called robust estimation methods developed in the past 15 years. These methods originally concerned the estimation of location parameters of a distribution from which a sample is drawn. [see Andrews et al. (1972), Hampel (1971, 1974), Huber (1964,1972)], but have recently been further developed for the linear regression case [see Andrews (1974), Forsythe (1972), Huber (1973) and Ronner (1977)]. In particular the class of robust M-estimators deserves attention. These estimators are of the maximum likelihood type, using possibly misspecified error densities. For the linear regression case this means that an estimate $\underline{b}_n$ of the parameter vector β_0 of the model $\underline{y}_j = \beta_0' x_j + \underline{u}_j$, $x_j \in R^k$, $j=1,2,\ldots$ is obtained by maximizing an objective function of the type $\underline{Q}_n^*(\beta) = \frac{1}{n}\sum_{j=1}^n \rho(\underline{y}_j - \beta' x_j)$, where ρ is a function such that $E\underline{Q}_n^*(\beta_0) = \max_\beta . E\underline{Q}_n^*(\beta)$. By choosing ρ such that $E\rho(\underline{u}_j + (\beta_0 - \beta)' x_j)$ is bounded we can avoid complications with infinite moments.

In this section we shall present a nonlinear robust M-estimation method. We shall adopt model (3.1.1) and assumption 3.1.1, but instead of assumption 3.1.2 we assume now:

<u>Assumption 3.2.1</u>. The error distribution $F(u)$ is symmetric unimodal with density $f(u)$.

For any symmetric unimodal density $\rho(u)$ we have by theorem 2.5.3 that

$$\psi(z) = \int_{-\infty}^{+\infty} \rho(u-z) f(u) du \tag{3.2.1}$$

is a density of a symmetric unimodal distribution. This implies that

$$E\,\rho(\underline{y}_j - g(\underline{x}_j,\theta)) = E\,\rho(\underline{u}_j - (g(\underline{x}_j,\theta) - g(\underline{x},\theta_0))) = E\,\psi(g(\underline{x}_j,\theta) - g(\underline{x}_j,\theta_0))$$

is maximal at $\theta = \theta_0$, and clearly the same applies for $E\,\frac{1}{n}\sum_{j=1}^n \rho(\underline{y}_j - g(\underline{x}_j,\theta))$. Intuitively we feel that the unique solution $\underline{\tilde{\theta}}_n$ obtained from

$$\frac{1}{n}\sum_{j=1}^n \rho(\underline{y}_j - g(\underline{x}_j,\underline{\tilde{\theta}}_n)) = \sup_{\theta\in\Theta} \frac{1}{n}\sum_{j=1}^n \rho(\underline{y}_j - g(\underline{x}_j,\theta))$$

can be used as an appropriate estimator of θ_0, as far as the left hand side is "almost" the mathematical expectation for large n.

If $\rho(u)$ is a symmetric unimodal density, so is $\frac{1}{\gamma}\rho(\frac{u}{\gamma})$ for $\gamma > 0$ and the same applies to

$$\phi(z|\gamma) = \int_{-\infty}^{\infty} \frac{1}{\gamma}\rho\left(\frac{u-z}{\gamma}\right) f(u) du\ . \tag{3.2.2}$$

Putting

$$\underline{R}_n(\theta|\gamma) = \frac{1}{n}\sum_{j=1}^n \frac{1}{\gamma}\rho\left(\frac{\underline{y}_j - g(\underline{x}_j,\theta)}{\gamma}\right)\quad,\ \gamma > 0\ , \tag{3.2.3}$$

we thus see that for any $\gamma > 0$

$$E\underline{R}_n(\theta|\gamma) = \frac{1}{n}\sum_{j=1}^{n} E\phi(g(\underline{x}_j,\theta) - g(\underline{x}_j,\theta_0)|\gamma)$$

is maximal at $\theta=\theta_0$. Moreover, if assumption 3.1.3 is satisfied and if ρ is continuous then from theorem 2.2.15

$$\lim E\underline{R}_n(\theta|\gamma) = \int\phi(g(x,\theta) - g(x,\theta_0)|\gamma)dH(x) , \tag{3.2.4}$$

which is also maximal at $\theta = \theta_0$ for any $\gamma > 0$. Thus also a solution $\tilde{\underline{\theta}}_n(\gamma)$ of

$$\underline{R}_n(\tilde{\underline{\theta}}_n(\gamma)|\gamma) = \sup_{\theta\in \Theta} \underline{R}_n(\theta|\gamma),\ \tilde{\underline{\theta}}_n(\gamma)\in\Theta\ ,\gamma > 0\ , \tag{3.2.5}$$

may be an appropriate estimator of θ_0. This statistic $\tilde{\underline{\theta}}_n(\gamma)$ will be called an M-estimator, and γ is the scaling parameter. In fact we now estimate θ_0 by a vector of random functions of the scaling parameter!

If the density ρ is chosen to be everywhere continuous, then it follows from lemma 3.1.2 that at least one of the solutions of (3.2.5) is, for given $\gamma>0$, a random vector, hence there exists a vector $\tilde{\underline{\theta}}_n(\gamma)$ of random functions satisfying (3.2.5). Therefore we assume throughout this chapter that ρ is everywhere continuous.

3.2.2 Strong consistency

Similar as in the case of least squares estimation, for showing that $\tilde{\underline{\theta}}_n(\gamma)\to\theta_0$ a.s. for $\gamma>0$ we have to show first that $\sup_{\theta\in\Theta} |\underline{R}_n(\theta|\gamma) - R(\theta|\gamma)| \to 0$ a.s. for $\gamma > 0$. However for later use we shall prove the more general result:

$$\sup_{(\theta,\gamma)\in\Theta\times\Gamma} |\underline{R}_n(\theta|\gamma) - R(\theta|\gamma)| \to 0 \quad \text{a.s.} \tag{3.2.6}$$

for any compact subinterval Γ of $(0,\infty)$. Second, we show that $R(\theta|\gamma)$ is a continuous function on Θ with a unique supremum on Θ at $\theta=\theta_0$.

If the density ρ is continuous everywhere, then it is also uniformly bounded. Thus (3.2.6) follows from the assumptions 3.1.1 and 3.1.3 and theorem 2.3.3, where

$$R(\theta|\gamma) = \int\phi(g(x,\theta_0) - g(x,\theta)|\gamma)dH(x) . \tag{3.2.7}$$

So it remains to show that $R(\theta|\gamma)$ has a unique supremum at $\theta=\theta_0$. For this we need the following lemma.

Lemma 3.2.1. Let $\rho(u)$ and $f(u)$ be symmetric unimodal densities and let $\rho(u)$ be strictly monotone increasing on $(-\infty,0)$. If $\rho(u)$ is differentiable and the derivative $\rho'(u)$ involved is uniformly bounded, then $\psi(z)=\int_{-\infty}^{+\infty}\rho(u-z)f(u)du$ is a symmetric unimodal density with unique mode at $z=0$.

Proof : Obviously $\psi(z)$ is symmetric unimodal. Since $\rho'(u)$ is uniformly bounded we may differentiate under the integral:

$$\psi'(z)=-\int_{-\infty}^{+\infty}\rho'(u-z)f(u)du \ .$$

Note that $\psi'(0)=0$ because $\rho'(u)$ is odd and $f(u)$ is even. Suppose that there is a $z_0 \neq 0$ such that $\psi'(z_0)=0$. Then

$$\psi'(z_0)=-\int_{-\infty}^{+\infty}\rho'(u-z_0)f(u)du=-\int_{-\infty}^{+\infty}\rho'(u)f(u+z_0)du=$$

$$=-\int_{-\infty}^{0}\rho'(u)\{f(u+z_0)-f(u-z_0)\}du=0 \ . \tag{3.2.8}$$

Without loss of generality we may assume $z_0>0$ because of the symmetry of ψ. Since $\rho'(u)>0$ and $f(u+z_0)\geq f(u-z_0)$ on $(-\infty,0)$,(3.2.8) implies that $f(u+z_0)=f(u-z_0)=0$; hence $f(u)=0$, on $(-\infty,z_0)$, except on a countable set D of discontinuities of $f(u)$. But from the unimodality of $f(u)$ it follows that $f(u)=0$ on $R^1\setminus D$ implies that $f(u)=0$ for $u\neq 0$, hence that the distribution involved is degenerated. Thus the lemma holds if the density $f(u)$ is not degenerated. However, otherwise $\psi(z)=\rho(z)$ and then the lemma is trivial. □

Thus under the conditions of the lemma we have for any $\gamma>0$

$$\phi(z|\gamma)<\phi(0|\gamma) \text{ if } z\neq 0 \ . \tag{3.2.9}$$

Now suppose that for some $\theta\neq\theta_0$ and some $\gamma>0$

$$\int\phi(g(z,\theta)-g(z,\theta_0)|\gamma)dH(x)=\phi(0|\gamma) \ .$$

Then (3.2.9) implies that the set $\{x\in R^p : g(x,\theta)\neq g(x,\theta_0)\}$ is a null set with respect to the distribution $H(x)$, and so is the set $\{x\in R^p : (g(x,\theta)-g(x,\theta_0))^2>0\}$. But then the simple functions $\chi(x)$, say, satisfying $0\leq\chi(x)\leq(g(x,\theta)-g(x,\theta_0))^2$ are all equal to zero for $x\in R^p$ and thus from definition 2.1.7 it follows that

$\int (g(x,\theta)- g(x,\theta_0))^2 dH(x)=0$, which contradicts assumption 3.1.5. Thus we have:

Lemma 3.2.2. If $\rho(u)$ satisfies the conditions of lemma 3.2.1 then under the assumptions 3.1.1, 3.2.1 and 3.1.5,

$$R(\theta|\gamma)= \int \phi(g(x,\theta)- g(x,\theta_0)|\gamma)dH(x)$$

has for every $\gamma > 0$ a unique supremum at $\theta=\theta_0$.

Now from (3.2.5), (3.2.6) and the lemma's 3.2.2 and 3.1.3 we conclude:

Theorem 3.2.1. Let $\rho(u)$ be a symmetric unimodal density with uniformly bounded derivative $\rho'(u)$ which is positive everywhere on $(-\infty,0)$. Then under the assumptions 3.1.1, 3.2.1, 3.1.3 and 3.1.5, $\tilde{\underline{\theta}}_n(\gamma) \to \theta_0$ a.s. for every $\gamma > 0$.

3.2.3 Asymptotic normality

We shall derive the limiting distribution of $\tilde{\underline{\theta}}_n(\gamma)$ in a similar way as for the least squares estimator. Thus again we consider the Taylor expansion

$$\sqrt{n}(\partial/\partial\theta)\underline{R}_n(\tilde{\underline{\theta}}_n(\gamma)|\gamma)= \sqrt{n}(\partial/\partial\theta)\underline{R}_n(\theta_0|\gamma)$$
$$+ \sqrt{n}(\tilde{\underline{\theta}}_n(\gamma)- \theta_0)'(\partial/\partial\theta)(\partial/\partial\theta')\underline{R}_n(\underline{\theta}_n(\gamma)|\gamma), \tag{3.2.10}$$

where:

$$|\underline{\theta}_n^*(\gamma)- \theta_0| \leq |\tilde{\underline{\theta}}_n(\gamma)-\theta_0|. \tag{3.2.11}$$

If $\rho'(u)$ and $\rho''(u)$ are continuous it follows, using lemma 3.1.4, that $\underline{\theta}_n^*(\gamma)$ is a random vector. Then (3.2.11) and theorem 3.2.1 imply

$$\underline{\theta}_n^*(\gamma) \to \theta_0 \quad \text{a.s. for every } \gamma > 0 \tag{3.2.12}$$

Furthermore, if ρ is twice differentiable and $\rho'(u)$ and $\rho''(u)$ are uniformly bounded, then we do not have to worry about the finiteness of the moments of the error distribution, as is easily seen by substituting (3.2.3) in (3.2.10). Similar as in section 3.1.3 it follows from lemma 3.1.5 and assumption 3.1.6 that:

$$\sqrt{n}(\partial/\partial\theta)\underline{R}_n(\tilde{\underline{\theta}}_n(\gamma)|\gamma) \to 0' \quad \text{a.s. for every } \gamma > 0. \tag{3.2.13}$$

Next consider an element of the matrix $(\partial/\partial\theta)(\partial/\partial\theta')\underline{R}_n(\theta|\gamma)$:

$$(\partial/\partial\theta_{i_1})(\partial/\partial\theta_{i_2})\underline{R}_n(\theta|\gamma)=$$

$$\frac{1}{n}\sum_{j=1}^{n}\frac{1}{\gamma^3}\rho''\left(\frac{\underline{u}_j+g(\underline{x}_j,\theta_0)-g(\underline{x}_j,\theta)}{\gamma}\right)\{(\partial/\partial\theta_{i_1})g(\underline{x}_j,\theta)\}\{(\partial/\partial\theta_{i_2})g(\underline{x}_j,\theta)\}$$

$$-\frac{1}{n}\sum_{j=1}^{n}\frac{1}{\gamma^2}\rho'\left(\frac{\underline{u}_j+g(\underline{x}_j,\theta_0)-g(\underline{x}_j,\theta)}{\gamma}\right)(\partial/\partial\theta_{i_1})(\partial/\partial\theta_{i_2})g(\underline{x}_j,\theta)$$

$$=\underline{a}^*_{n,i_1,i_2}(\theta|\gamma)-\underline{b}^*_{n,i_1,i_2}(\theta|\gamma), \tag{3.2.14}$$

say. Similar to (3.1.21) it follows that from the assumptions 3.1.1, 3.1.3 and 3.1.7 and theorem 2.3.3 :

$$\sup_{(\theta,\gamma)\in\Theta\times\Gamma}|\underline{a}^*_{n,i_1,i_2}(\theta|\gamma)-a^*_{i_1,i_2}(\theta|\gamma)|\to 0 \quad \text{a.s.} \tag{3.2.15}$$

for any compact subinterval Γ of $(0,\infty)$, where

$$a^*_{i_1,i_2}(\theta|\gamma)=\iint\frac{1}{\gamma^3}\rho''\left(\frac{u+g(x,\theta_0)-g(x,\theta)}{\gamma}\right)f(u)du\{(\partial/\partial\theta_{i_1})g(x,\theta)\}\{(\partial/\partial\theta_{i_2})g(x,\theta)\}dH(x). \tag{3.2.16}$$

Hence by (3.2.12) and theorem 2.3.2

$$\underline{a}^*_{n,i_1,i_2}(\underline{\theta}_n(\gamma)|\gamma)\to a^*_{i_1,i_2}(\theta_0|\gamma)=\int\frac{1}{\gamma^2}\rho''(\frac{u}{\gamma})f(u)du.a_{i_1,i_2}(\theta_0) \text{ a.s.}, \tag{3.2.17}$$

where $a_{i_1,i_2}(\theta)$ is defined by (3.1.22).

If assumption 3.1.8 is weakened to

<u>Assumption 3.2.2.</u> $\sup_n\frac{1}{n}\sum_{j=1}^{n}E\left(\overline{(\partial/\partial\theta_{i_1})(\partial/\partial\theta_{i_2})g(\underline{x}_j,\theta)}^{\Theta}\right)^{1+\delta}<\infty$

for $i_1,i_2=1,2,\ldots q$, and some $\delta>0$,

then, using inequality(2.3.17) and the fact that ρ' is uniformly bounded,

$$\sup_n\frac{1}{n}\sum_{j=1}^{n}E\left\{\overline{\frac{1}{\gamma^2}\rho'\left(\frac{\underline{u}_j+g(\underline{x}_j,\theta_0)-g(\underline{x}_j,\theta_0)}{\gamma}\right)(\partial/\partial\theta_{i_1})(\partial/\partial\theta_{i_2})g(\underline{x}_j,\theta)}^{\Theta\times\Gamma}\right\}^{1+\delta}<\infty$$

for any compact subinterval Γ of $(0,\infty)$, hence from theorem 2.3.4

$$\text{plim}\sup_{(\theta,\gamma)\in\Theta\times\Gamma}|\underline{b}^*_{n,i_1,i_2}(\theta|\gamma)-b^*_{i_1,i_2}(\theta|\gamma)|=0 . \tag{3.2.19}$$

Consequently by (3.2.12) and theorem 2.3.2

$$\text{plim}\ \underline{b}^*_{n,i_1,i_2}(\underline{\theta}^*_n(\gamma)|\gamma)=b^*_{i_1,i_2}(\theta_0|\gamma)=$$

$$=-\frac{1}{\gamma 2}\int \rho'(\frac{u}{\gamma})f(u)du\int(\partial/\partial\theta_{i_1})(\partial/\partial\theta_{i_2})g(x,\theta_0)dH(x)=0, \qquad (3.2.20)$$

where the latter conclusion follows from the fact that the left side integral is zero because ρ' is odd and f is symmetric. Thus similar to lemma 3.1.6 we now have:

Lemma 3.2.3. Under the assumptions of theorem 3.2.1 and the additional assumptions 3.1.6, 3.1.7 and 3.2.2, we have
$(\partial/\partial\theta)(\partial/\partial\theta')\underline{R}_n(\underline{\theta}^*_n(\gamma)|\gamma)\to\frac{1}{\gamma 3}\int\rho''(\frac{u}{\gamma})f(u)du.A(\theta_0)$ in pr.,

provided that $\rho'(u)$ and $\rho''(u)$ are continuous and uniformly bounded,

where $A(\theta)$ is defined in lemma 3.1.6. From the argument in section 3.1.3 we have seen that for deriving the limiting distribution of $\sqrt{n}(\underline{\tilde{\theta}}_n(\gamma)-\theta_0)$ from the Taylor expansion (3.2.10) it is necessary that the limit matrix in the lemma above is non singular. Adopting assumption 3.1.9 we thus need a further condition for assuring $\int\rho''(\frac{u}{\gamma})f(u)du \neq 0$. This further condition may be:

Assumption 3.2.3. The density f(u) is everywhere differentiable,

because then, together with assumption 3.2.1 it follows that $f'(u)$ is non negative on $(-\infty,0)$ and non positive on $(0,\infty)$, and hence $\rho'(\frac{u}{\gamma})f'(u)$ is non negative on $(-\infty,0)$ and $(0,\infty)$, provided that ρ is chosen as in theorem 3.2.1. Since

$\int\rho''(\frac{u}{\gamma})f(u)du=-\gamma\int\rho'(\frac{u}{\gamma})f'(u)du$ and $\rho'(\frac{u}{\gamma})f'(u)$ cannot be zero everywhere because of the unimodality of both ρ and f, we thus have:

Lemma 3.2.4. If ρ is chosen such as in theorem 3.2.1 then under the assumptions 3.2.1 and 3.2.2, $\int\rho''(\frac{u}{\gamma})f(u)du < 0$ for all $\gamma > 0$.

The proof of the asymptotic normality of $\sqrt{n}(\partial/\partial\theta')\underline{R}_n(\theta_0|\gamma)$ is similar to the proof of (3.1.31), provided that $\rho'(u)$ is uniformly bounded. Thus we conclude

Lemma 3.2.5. Under the assumptions 3.1.1, 3.1.7 and 3.1.9,
$\sqrt{n}(\partial/\partial\theta')\underline{R}_n(\theta_0|\gamma)\to N_q[0,\int\frac{1}{\gamma^4}\rho'(\frac{u}{\gamma})^2f(u)du\ A(\theta_0)]$ in distr.

for every $\gamma > 0$, provided that $\rho'(u)$ is uniformly bounded.

Combining the results of the lemma's 3.2.3, 3.2.4 and 3.2.5 we obtain:

Theorem 3.2.2. If $\rho'(u)$ and $\rho''(u)$are continuous and uniformly bounded then under the assumptions of theorem 3.2.1 and the additional assumptions 3.1.6, 3.1.7, 3.2.2, 3.2.3 and 3.1.9,

$$\sqrt{n}(\underline{\tilde{\theta}}_n(\gamma)-\theta_0) \to N_q[0,\ h(\gamma)A(\theta_0)^{-1}] \text{ in distr.}$$

for every $\gamma > 0$, where

$$h(\gamma)=\frac{h_1(\gamma)}{h_2(\gamma)^2}=\frac{\gamma^2\int\rho'(\frac{u}{\gamma})^2 f(u)du}{\{\int\rho''(\frac{u}{\gamma})f(u)du\}^2} > 0 \qquad (3.2.21)$$

for every $\gamma > 0$.

From this theorem it is now clear why we have taken the parameter γ into account; namely because the asymptotic efficiency of $\underline{\tilde{\theta}}_n(\gamma)$ depends on it. Furthermore, since we have not used any assumption about the moments of the error distribution, the theorems 3.2.1 and 3.2.2 still hold if for example the error distribution is symmetric stable with characteristic exponent $0<\alpha<2$.

3.2.4 Properties of the function $h(\gamma)$. Asymptotic efficiency and robustness

Let us have a closer look at the function $h(\gamma)$. Since $\frac{\rho'(\frac{u}{\gamma})}{\frac{u}{\gamma}}=\frac{\rho'(\frac{u}{\gamma})-\rho'(0)}{\frac{u}{\gamma}} \to \rho''(0)$

if $\gamma \to \infty$, it follows from the bounded convergence theorem (theorem 2.2.7) that

$$h_1(\gamma)=\gamma^2\int\rho'(\tfrac{u}{\gamma})^2 f(u)du=\int u^2\left\{\frac{\rho'(\frac{u}{\gamma})}{\frac{u}{\gamma}}\right\}^2 f(u)du \to \rho''(0)^2\int u^2 f(u)du \text{ as } \gamma\to\infty,$$

provided that $\int u^2 f(u)du < \infty$. But if $\int u^2 f(u)du = \infty$ then from Fatou's lemma (theorem 2.2.8)

$$\infty=\int \liminf_{\gamma\to\infty}.\ u^2\left\{\frac{\rho'(\frac{u}{\gamma})}{\frac{u}{\gamma}}\right\}^2 f(u)du=\int u^2\rho''(0)^2 f(u)du \le \lim_{\gamma\to\infty} h_1(\gamma).$$

Thus

$$\lim_{\gamma\to\infty} h_1(\gamma)=\rho''(0)^2\int u^2 f(u)du \le \infty . \qquad (3.2.22)$$

Moreover, again from the bounded convergence theorem we have

$$\lim_{\gamma\to\infty} h_2(\gamma)= \int \lim_{\gamma\to\infty} \rho''(\tfrac{u}{\gamma})f(u)du=\rho''(0) \qquad (3.2.23)$$

and consequently:

<u>Theorem 3.2.3</u>. If $\rho''(0)\neq 0$ then under the assumptions of theorem 3.2.2, $\lim_{\gamma\to\infty} h(\gamma)= \int u^2 f(u)du$, where the limit involved may be infinite.

Thus if the conditions for consistency and asymptotic normality of the least squares estimator are satisfied then the asymptotic efficiency of our M-estimator $\tilde{\underline{\theta}}_n(\gamma)$ approaches that of the least squares estimator if γ is chosen large, but if assumption 3.1.2 fails to hold then $\tilde{\underline{\theta}}_n(\gamma)$ is more efficient. Therefore our M-estimator is <u>robust</u>.

Suppose that $\int u^2 f(u)du < \infty$ and choose $\rho(u)= \frac{e^{-\frac{1}{2}u^2}}{\sqrt{2\pi}}$. Then

$$h_1(\gamma)=\gamma^2 \int \rho'(\tfrac{u}{\gamma})^2 f(u)du= \frac{1}{2\pi} \int u^2 e^{-u^2/\gamma^2} f(u)du \qquad (3.2.24)$$

$$h_2(\gamma)=\int \rho''(\tfrac{u}{\gamma}) f(u)du=\frac{1}{\sqrt{2\pi}}\int (\tfrac{u}{\gamma})^2 e^{-\frac{1}{2}u^2/\gamma^2} f(u)du- \frac{1}{\sqrt{2\pi}} \int e^{-\frac{1}{2}u^2/\gamma^2} f(u)du , \qquad (3.2.25)$$

$$h_1'(\gamma)= \frac{1}{\pi\gamma^3} \int u^4 e^{-u^2/\gamma^2} f(u)du , \qquad (3.2.26)$$

$$h_2'(\gamma)= \frac{-3}{\gamma^3\sqrt{2\pi}}\int u^2 e^{-\frac{1}{2}u^2/\gamma^2} f(u)du + \frac{1}{\gamma^3\sqrt{2\pi}}\int u^2(\tfrac{u}{\gamma})^2 e^{-\frac{1}{2}u^2/\gamma^2} f(u)du . \qquad (3.2.27)$$

From the bounded convergence theorem we have

$$\lim_{\gamma\to\infty} h_1(\gamma)= \frac{1}{2\pi} \int u^2 f(u)du ,$$

$$\lim_{\gamma\to\infty} h_2(\gamma)=- \frac{1}{\sqrt{2\pi}}$$

and

$$\lim_{\gamma\to\infty} \gamma^3 h_2'(\gamma)= - \frac{3}{\sqrt{2\pi}} \int u^2 f(u)du ,$$

while from the monotone convergence theorem

$$\lim_{\gamma\to\infty} \gamma^3 h_1'(\gamma)= \frac{1}{\pi} \int u^4 f(u)du.$$

Thus

$$\lim_{\gamma\to\infty} \gamma^3 h'(\gamma) = \lim_{\gamma\to\infty} \frac{\gamma^3 h_1'(\gamma)h_2(\gamma) - 2h_1(\gamma)\gamma^3 h_2'(\gamma)}{h_2(\gamma)^3}$$

$$= 2\{\int u^4 f(u)du - 3(\int u^2 f(u)du)^2\} \qquad (3.2.28)$$

and consequently, if the <u>kurtosis</u>:

$$\int_{-\infty}^{+\infty} u^4 f(u)du\{\int_{-\infty}^{+\infty} u^2 f(u)du\}^{-2} - 3$$

of the error density f is positive (or ∞), or with other words: if the error density is <u>leptokurtic</u>, then $\lim_{\gamma\to\infty} h(\gamma) = \int u^2 f(u)du$ is approached from below. In its turn this implies that there is a $\gamma_* > 0$ such that $h(\gamma) < \int u^2 f(u)du$ for all $\gamma \in (\gamma_*, \infty)$.

This result is not specific for $\rho(u) = \frac{e^{-\frac{1}{2}u^2}}{\sqrt{2\pi}}$. It is easy to verify that the same applies if we choose for example

$$\rho(u) = \left(\frac{1}{1+u^2}\right)^\beta \Big/ \int_{-\infty}^{+\infty} \left(\frac{1}{1+u^2}\right)^\beta du \quad , \beta \geq 1. \qquad (3.2.29)$$

(Note that for $\beta=1$, (3.2.29) is just the symmetric Cauchy density)

<u>Theorem 3.2.4</u>. If $\rho(u)$ is chosen to be a symmetric normal density or a density of the type (3.2.29) and if

$$\int u^2 f(u)du < \infty, \int u^4 f(u)du\{\int u^2 f(u)du\}^{-2} - 3 > 0 , \qquad (3.2.30)$$

then there is a number $\gamma_* > 0$ such that $h(\gamma) < \int u^2 f(u)du$ for all $\gamma \in (\gamma_*, \infty)$.

This theorem indicates that even if all the conditions for consistency and asymptotic normality of the least squares estimator are satisfied, our M-estimator may be more efficient. For example this is the case if the error distribution is a <u>mixture</u> of two normal distribution $N(0,\sigma_1^2)$ and $N(0,\sigma_2^2)$ with $\sigma_1^2 \neq \sigma_2^2$ because such mixtures are leptokurtic, while obviously they have finite variances. However, if the error distribution is normal, then the kurtosis equals zero, as is well known, so that then theorem 3.2.4 does not apply. But this is not supprising, because then the least squares estimator is also a maximum likelihood estimator.

Next we question what happens with $h(\gamma)$ if we let $\gamma\downarrow 0$. First we observe that

$$h_1(\gamma)=\gamma^2\int\rho'(\tfrac{u}{\gamma})^2 f(u)du=\gamma^3\int\rho'(u)^2 f(\gamma u)du$$

and hence from the bounded convergence theorem,

$$\lim_{\gamma\downarrow 0}\frac{h_1(\gamma)}{\gamma^3}=f(0)\int\rho'(u)^2du\ , \tag{3.2.31}$$

provided that f is continuous at 0. Second, if f is twice differentiable then

$$h_2(\gamma)=\int\rho''(\tfrac{u}{\gamma})f(u)du=\gamma\int\rho''(u)f(\gamma u)du=-\gamma^2\int\rho'(u)f'(\gamma u)du$$

$$=\gamma^3\int\rho(u)f''(\gamma u)du$$

and hence

$$\lim_{\gamma\downarrow 0}\frac{h_2(\gamma)}{\gamma^3}=f''(0)\ , \tag{3.2.32}$$

provided that f'' is continuous at 0. Consequently we now have:

Theorem 3.2.5. If f is twice differentiable and if f'' is continuous and non-zero at 0, then

$$\lim_{\gamma\downarrow 0}\gamma^3 h(\gamma)=\frac{f(0)}{(f''(0))^2}\int\rho'(u)^2du$$

and consequently

$$\lim_{\gamma\downarrow 0}h(\gamma)=\infty\ .$$

3.2.5 A uniformly consistent estimator of the function $h(\gamma)$

We will now be concerned with the problem of estimating the function $h(\gamma)$. First, we define

$$\underline{h}_{1,n}(\gamma,\theta)=\frac{1}{n}\sum_{j=1}^n\gamma^2\rho'\left(\frac{\underline{y}_j-g(\underline{x}_j,\theta)}{\gamma}\right)^2$$

$$=\frac{1}{n}\sum_{j=1}^n\gamma^2\rho'\left(\frac{\underline{u}_j+g(\underline{x}_j,\theta_0)-g(\underline{x}_j,\theta)}{\gamma}\right)^2 \tag{3.2.33}$$

Since $\rho'(u)$ is continuous and uniformly bounded, it follows similar to (3.2.6) that under the assumptions 3.1.1 and 3.1.3

$$\sup_{(\theta,\gamma)\in \text{Ⓗ} \times \Gamma} |\underline{h}_{1,n}(\gamma,\theta) - h_1(\gamma,\theta)| \to 0 \quad \text{a.s.} \tag{3.2.34}$$

for any compact subinterval Γ of $(0,\infty)$, where

$$h_1(\gamma,\theta)=\gamma^2\iint \rho'\left(\frac{u + g(x,\theta) - g(x,\theta_0)}{\gamma}\right)^2 f(u)du\, dH(x) . \tag{3.2.35}$$

Next, putting

$$\underline{h}_{2,n}(\gamma,\theta)= \frac{1}{n} \sum_{j=1}^{n}\rho''\left(\frac{\underline{y}_j - g(\underline{x}_j,\theta)}{\gamma}\right) , \tag{3.2.36}$$

it follows similarly that

$$\sup_{(\theta,\gamma)\in \text{Ⓗ} \times \Gamma}\left|\underline{h}_{2,n}(\gamma,\theta)- h_2(\gamma,\theta)\right| \to 0 \quad \text{a.s.} \quad , \tag{3.2.37}$$

where

$$h_2(\gamma,\theta)= \iint\rho''\left(\frac{u + g(x,\theta_0)- g(x,\theta)}{\gamma}\right)f(u)du\, dH(x) . \tag{3.2.38}$$

Since $h_1(\gamma,\theta)$ and $h_2(\gamma,\theta)$ are continuous on $\Gamma \times$ Ⓗ and Γ is compact, it follows from theorem 2.3.1 that $\sup_{\gamma\in\Gamma}|h_1(\gamma,\theta)- h_1(\gamma,\theta_0)|$ and $\sup_{\gamma\in\Gamma}|h_2(\gamma,\theta)- h_2(\gamma,\theta_0)|$ are continuous on Ⓗ , and hence also uniformly continuous on Ⓗ because Ⓗ is compact. But then it follows from theorem 2.2.5 that under the condition of theorem 3.2.1,

$$\sup_{\gamma\in\Gamma}|h_1(\gamma,\underline{\tilde{\theta}}_n(\gamma_1))- h_1(\gamma,\theta_0)| \to 0 \qquad \text{a.s.} \tag{3.2.39}$$

$$\sup_{\gamma\in\Gamma}|h_2(\gamma,\underline{\tilde{\theta}}_n(\gamma_1))- h_2(\gamma,\theta_0)| \to 0 \qquad \text{a.s.} \tag{3.2.40}$$

for any $\gamma_1 > 0$.

Since (3.2.34) and (3.2.37) imply

$$\sup_{\gamma\in\Gamma}|\underline{h}_{1,n}(\gamma,\underline{\tilde{\theta}}_n(\gamma_1))- h_1(\gamma,\underline{\tilde{\theta}}_n(\gamma_1))| \to 0 \quad \text{a.s.} \tag{3.2.41}$$

and

$$\sup_{\gamma\in\Gamma}|\underline{h}_{2,n}(\gamma,\underline{\tilde{\theta}}_n(\gamma_1)) - h_2(\gamma,\underline{\tilde{\theta}}_n(\gamma_1))|\to 0 \quad \text{a.s.} \tag{3.2.42}$$

and since $h_1(\gamma,\theta_0) = h_1(\gamma)$ and $h_2(\gamma,\theta_0) = h_2(\gamma)$, we thus have

$$\sup_{\gamma\in\Gamma} |\underline{h}_{1,n}(\gamma,\underline{\tilde{\theta}}_n(\gamma_1)) - h_1(\gamma)| \to 0 \text{ a.s. for every } \gamma_1 > 0 , \tag{3.2.43}$$

$$\sup_{\gamma\in\Gamma} |\underline{h}_{2,n}(\gamma,\underline{\tilde{\theta}}_n(\gamma_1)) - h_2(\gamma)| \to 0 \text{ a.s. for every } \gamma_1 > 0 . \tag{3.2.44}$$

Moreover, since $\underline{h}_{2,n}(\gamma,\theta)$ and $h_2(\gamma)$ are bounded on $\Gamma \times Ⓗ$ and Γ, respectively, (3.2.44) implies:

$$\sup_{\gamma\in\Gamma} |\underline{h}_{2,n}(\gamma,\underline{\tilde{\theta}}_n(\gamma_1))^2 - h_2(\gamma)^2| \to 0 \text{ a.s. for every } \gamma_1 > 0 \tag{3.2.45}$$

while from (3.2.21) and lemma 3.2.4 we have

$$\inf_{\gamma\in\Gamma} |h_2(\gamma)^2| > 0 . \tag{3.2.46}$$

Of course, from the continuity of $h_1(\gamma)$ and $h_2(\gamma)$ and the compactness of Γ we also have

$$\sup_{\gamma\in\Gamma} |h_1(\gamma)| < \infty, \quad \sup_{\gamma\in\Gamma} |h_2(\gamma)| < \infty . \tag{3.2.47}$$

Putting

$$\underline{h}_n(\gamma,\theta) = \frac{\underline{h}_{1,n}(\gamma,\theta)}{\underline{h}_{2,n}(\gamma,\theta)^2} , \tag{3.2.48}$$

we now see from the trivial equality

$$\frac{x_*}{y_*^2} - \frac{x}{y^2} = \frac{(x_* - x) + x}{(y_*^2 - y^2) + y^2} - \frac{x}{y^2} = \frac{y^2(x_* - x) - x(y_*^2 - y^2)}{y^4 + y^2(y_*^2 - y^2)}$$

that

$$\sup_{\gamma\in\Gamma} |\underline{h}_n(\gamma,\underline{\tilde{\theta}}_n(\gamma_1)) - h(\gamma)| \le$$

$$\frac{\sup_{\gamma\in\Gamma} h_2(\gamma)^2 \sup_{\gamma\in\Gamma} |\underline{h}_{1,n}(\gamma,\underline{\tilde{\theta}}_n(\gamma_1)) - h_1(\gamma)| + \sup_{\gamma\in\Gamma} |h_1(\gamma)| \sup_{\gamma\in\Gamma} |\underline{h}_{2,n}(\gamma,\underline{\tilde{\theta}}_n(\gamma_1))^2 - h_2(\gamma)^2|}{\inf_{\gamma\in\Gamma} h_2(\gamma)^4 - \sup_{\gamma\in\Gamma} h_2(\gamma)^2 \sup_{\gamma\in\Gamma} |h_{2,n}(\gamma,\underline{\tilde{\theta}}_n(\gamma_1))^2 - h_2(\gamma)^2|} . \tag{3.2.49}$$

Hence from (3.2.43) and (3.2.45) through (3.2.49) we now have

$$\sup_{\gamma\in\Gamma} |\underline{h}_n(\gamma,\underline{\tilde{\theta}}_n(\gamma_1)) - h(\gamma)| \to 0 \text{ a.s. for any } \gamma_1 > 0 . \tag{3.2.50}$$

If $\int u^2 f(u)du < \infty$, then it follows from theorem 3.2.3 that $h(\frac{1}{\lambda})$ is continuous and uniformly bounded on $[0,\lambda_*]$ for any $\lambda_* > 0$ (provided $\rho''(0) \neq 0$). By substituting $\gamma = \frac{1}{\lambda}$ we then see that

$$\lim_{\lambda\downarrow 0} \underline{h}_{1,n}(\tfrac{1}{\lambda},\theta) = \frac{1}{n}\sum_{j=1}^{n}\left\{\lim_{\lambda\downarrow 0}\frac{\rho'(\lambda(\underline{y}_j - g(\underline{x}_j,\theta))) - \rho'(0)}{\lambda}\right\}^2$$

$$= \rho''(0)^2\,\frac{1}{n}\sum_{j=1}^{n}(\underline{y}_j - g(\underline{x}_j,\theta))^2 \qquad (3.2.51)$$

and

$$\lim_{\lambda\downarrow 0} \underline{h}_{2,n}(\tfrac{1}{\lambda},\theta) = \frac{1}{n}\sum_{j=1}^{n}\lim_{\lambda\downarrow 0}\rho''(\lambda(\underline{y}_j - g(\underline{x}_j,\theta))) = \rho''(0)\ , \qquad (3.2.52)$$

hence

$$\lim_{\lambda\downarrow 0} \underline{h}_n(\tfrac{1}{\lambda},\theta) = \underline{Q}_n(\theta)\ ,$$

where $\underline{Q}_n(\theta)$ is defined by (3.1.3). Thus if $\int u^2 f(u)du < \infty$ and $\rho''(0) \neq 0$ then (3.2.50) carries over for $\Gamma = [\gamma_*,\infty)$ with $\gamma_* > 0$. Summarizing:

<u>Theorem 3.2.6.</u> Under the assumptions of theorem 3.2.2 we have $\sup_{\gamma\in\Gamma}|\underline{h}_n(\gamma,\underline{\tilde{\theta}}_n(\gamma_1)) - h(\gamma)| \to 0$ a.s. for any $\gamma_1 > 0$ and any compact subinterval Γ of $(0,\infty)$. Moreover, if $\int u^2 f(u)du < \infty$ and $\rho''(0) \neq 0$, then we may choose $\Gamma = [\gamma_*,\infty)$ with $\gamma_* > 0$.

<u>Remark:</u> The result of theorem 3.2.5, i.e. $\lim_{\gamma\downarrow 0} h(\gamma) = \infty$, does not automatically carry over to the estimator $\underline{h}_n(\gamma,\underline{\hat{\theta}}_n(\gamma_1))$. For example, chosing $\rho(u) = \exp(-\frac{1}{2}u^2)/\sqrt{2\pi}$, we have:

$$\underline{h}_n(\gamma,\underline{\tilde{\theta}}_n(\gamma_1)) = \frac{\underline{h}_{1,n}(\gamma,\underline{\tilde{\theta}}_n(\gamma_1))}{h_{2,n}(\gamma,\underline{\tilde{\theta}}_n(\gamma_1))^2} = \frac{\frac{1}{n}\sum_{j=1}^{n}\underline{\hat{u}}_j^2\exp(-\underline{\hat{u}}_j^2/\gamma^2)}{\{\frac{1}{n}\sum_{j=1}^{n}(\underline{\hat{u}}_j^2/\gamma^2 - 1)\exp(-\frac{1}{2}\underline{\hat{u}}_j^2/\gamma^2)\}^2}$$

$$\leq \frac{\frac{1}{n}\sum_{j=1}^{n}\underline{\hat{u}}_j^2\exp(-\min\underline{\hat{u}}_j^2/\gamma^2)}{\{\frac{1}{n}\sum_{j=1}^{n}\left(\frac{\underline{\hat{u}}_j^2}{\gamma^2} - 1\right)\exp(-\frac{1}{2}\underline{\hat{u}}_j^2/\gamma^2)\}^2}$$

$$= \frac{\frac{1}{n}\sum_{j=1}^{n}\underline{\hat{u}}_j^2}{\{\frac{1}{n}\sum_{j=1}^{n}\left(\frac{\underline{\hat{u}}_j^2}{\gamma^2} - 1\right)\exp(-\frac{1}{2}(\underline{\hat{u}}_j^2 - \min\underline{\hat{u}}_j^2)/\gamma^2)\}^2}$$

$$= \frac{\frac{1}{n}\sum_{j=1}^{n}\underline{\hat{u}}_j^2}{\{\sum_{\underline{\hat{u}}_j^2=\min\underline{\hat{u}}_j^2}\left(\frac{\min\underline{\hat{u}}_j^2}{\gamma^2} - 1\right) + \sum_{\underline{\hat{u}}_j^2>\min\underline{\hat{u}}_j^2}\left(\frac{\underline{\hat{u}}_j^2}{\gamma^2} - 1\right)\exp(-\frac{1}{2}(\underline{\hat{u}}_j^2 - \min\underline{\hat{u}}_j^2)/\gamma^2)\}^2}$$

$$\rightarrow \begin{cases} \frac{1}{n}\sum_{j=1}^{n}\underline{\hat{u}}_j^2/(\sum_{\underline{\hat{u}}_j=0} 1)^2 & \text{if} \quad \min \underline{\hat{u}}_j^2=0 \quad \text{and} \quad \gamma\downarrow 0 \\ 0 & \text{if} \quad \min \underline{\hat{u}}_j^2>0 \quad \text{and} \quad \gamma\downarrow 0 \end{cases},$$

where the $\underline{\hat{u}}_j$'s are the residuals: $\underline{\hat{u}}_j=\underline{y}_j-g(\underline{x}_j,\underline{\tilde{\theta}}_n(\gamma_1))$. Thus γ_* should not be chosen too small! But how small is "too small"? From an asymptotic point of view any $\gamma_* > 0$ is appropriate, but this is not the case for small samples. For solving this problem, consider the result (3.2.44) and notice that from lemma 3.2.4, $h_2(\gamma) < 0$ for all $\gamma > 0$. Thus $\underline{h}_{2,n}(\gamma,\underline{\tilde{\theta}}_n(\gamma_1))$ converges a.e. to a negative valued function. This suggest that γ_* should be chosen such that

$$\sup_{\gamma\in[\gamma_*,\infty)} \underline{h}_{2,n}(\gamma,\underline{\tilde{\theta}}_n(\gamma_1)) < 0 \ .$$

3.2.6 A two stage robust M-estimator

Now let Γ be compact and suppose that there is a unique point γ_0 in Γ such that:

$$h(\gamma_0)= \inf_{\gamma\in\Gamma} h(\gamma) \ . \tag{3.2.53}$$

Moreover, let $\underline{\tilde{\gamma}}_n(\gamma_1)$ be such that

$$\underline{h}_n(\underline{\tilde{\gamma}}_n(\gamma_1),\ \underline{\tilde{\theta}}_n(\gamma_1))= \inf_{\gamma\in\Gamma} \underline{h}_n(\gamma,\ \underline{\tilde{\theta}}_n(\gamma_1)), \qquad \underline{\tilde{\gamma}}_n(\gamma_1)\in\Gamma \ . \tag{3.2.54}$$

Then it follows from theorem 3.2.6 and lemma 3.1.3 that

$$\underline{\tilde{\gamma}}_n(\gamma_1)\rightarrow \gamma_0 \quad \text{a.s. for any } \gamma_1 > 0 \ . \tag{3.2.55}$$

If we substitute $\gamma = \underline{\tilde{\gamma}}_n(\gamma_1)$ in (3.2.5) we then get a two stage M-estimator $\underline{\tilde{\theta}}_n(\underline{\tilde{\gamma}}_n(\gamma_1))$ which is also strongly consistent:

$$\underline{\tilde{\theta}}_n(\underline{\tilde{\gamma}}_n(\gamma_1)) \rightarrow \theta_0 \quad \text{a.s. for any } \gamma_1 > 0, \tag{3.2.56}$$

because from (3.2.6) we have

$$\sup_{\theta\in \textcircled{H}} |\underline{R}_n(\theta|\underline{\tilde{\gamma}}_n(\gamma_1))- R(\theta|\tilde{\gamma}(\gamma_1))|\rightarrow 0 \ , \tag{3.2.57}$$

while from (3.2.55), and the uniform continuity of $R(\theta|\gamma)$ on $\textcircled{H} \times \Gamma$ it follows that:

$$\sup_{\theta\in \textcircled{H}} |R(\theta|\tilde{\gamma}_n(\gamma_1))- R(\theta|\gamma_0)|\rightarrow 0 \text{ a.s.} \ , \tag{3.2.58}$$

so that:

$$\sup_{\theta\in\textcircled{H}} |R_n(\theta|\tilde{\gamma}_n(\gamma_1)) - R(\theta|\gamma_0)| \to 0 \quad \text{a.s.}\ , \tag{3.2.59}$$

which by lemma 3.1.3 implies (3.2.56).

Next we show that the two stage M-estimator $\tilde{\underline{\theta}}_n(\tilde{\underline{\gamma}}_n(\gamma_1))$ has the same asymptotic properties as $\tilde{\underline{\theta}}_n(\gamma_0)$.

It follows from (3.2.55), (3.2.15) and (3.2.19) that for any $\gamma_1 > 0$,

$$\sup_{\theta\in\textcircled{H}} |\underline{a}^*_{n,i_1,i_2}(\theta|\tilde{\underline{\gamma}}_n(\gamma_1)) \to a^*_{i_1,i_2}(\theta|\gamma_0)| \to 0 \quad \text{a.s.} \tag{3.2.60}$$

and

$$\sup_{\theta\in\textcircled{H}} |\underline{b}^*_{n,i_1,i_2}(\theta|\tilde{\underline{\gamma}}_n(\gamma_1)) - b^*_{i_1,i_2}(\theta|\gamma_0)| \to 0 \quad \text{in pr..} \tag{3.2.61}$$

Hence (compare lemma 3.2.3),

$$(\partial/\partial\theta)(\partial/\partial\theta')\underline{R}_n(\underline{\theta}^*_n(\tilde{\underline{\gamma}}_n(\gamma_1))|\tilde{\underline{\gamma}}_n(\gamma_1)) \to \frac{1}{\gamma_0^3}\int\rho''(\frac{u}{\gamma_0})f(u)du A(\theta_0) \quad \text{in pr..} \tag{3.2.62}$$

Furthermore, obviously also (3.2.13) carries over:

$$\sqrt{n}(\partial/\partial\theta)\underline{R}_n(\tilde{\underline{\theta}}_n(\tilde{\underline{\gamma}}_n(\gamma_1))|\ \tilde{\underline{\gamma}}_n(\gamma_1)) \to 0' \quad \text{a.s. for any } \gamma_1 > 0, \tag{3.2.63}$$

provided that assumption 3.1.6 is satisfied.

But what about lemma 3.2.5? We show now that also this lemma carries over, provided that ρ is suitable chosen. Thus we consider

$$(\partial/\partial\theta_i)\underline{R}_n(\theta_0|\gamma) = -\frac{1}{n}\sum_{j=1}^n \frac{1}{\gamma^2}\rho'(\frac{\underline{u}_j}{\gamma})(\partial/\partial\theta_i)g(\underline{x}_j,\theta_0). \tag{3.2.64}$$

From theorem 2.4.5 and the symmetry of the density $\rho(u)$ it follows that

$$\rho(u) = \frac{1}{2\pi}\int\cos(tu)\eta(t)dt \tag{3.2.65}$$

and consequently

$$\rho'(u) = -\frac{1}{2\pi}\int\sin(tu)t\eta(t)dt, \tag{3.2.66}$$

provided that

$$\int|t||\eta(t)|dt < \infty, \tag{3.2.67}$$

where $\eta(t)$ is the characteristic function of $\rho(u)$. From (3.2.66) we now obtain:

$$\frac{1}{\gamma^2}\rho'(\frac{u}{\gamma}) = -\frac{1}{2\pi}\int \sin(t\frac{u}{\gamma})\frac{t}{\gamma}\eta(t)d(\frac{t}{\gamma}) = -\frac{1}{2\pi}\int \sin(tu)t\eta(\gamma t)dt. \quad (3.2.68)$$

Substituting (3.2.68) in (3.2.64) yields

$$\underline{v}_{i,n}(\gamma) = \sqrt{n}(\partial/\partial\theta_i)\underline{R}_n(\theta_0|\gamma) - \sqrt{n}(\partial/\partial\theta_i)\underline{R}_n(\theta_0|\gamma_0)$$

$$= -\frac{1}{2\pi\gamma_0^2}\int\{\frac{1}{\sqrt{n}}\sum_{j=1}^n \sin(t\underline{u}_j)(\partial/\partial\theta_i)g(\underline{x}_j,\theta_0)\}t\{\eta(\gamma t) - \eta(\gamma_0 t)\}dt$$

$$= -\frac{1}{2\pi\gamma_0^2}\int \underline{w}_{i,n}(t)(\frac{\eta(\gamma t)}{\eta(\gamma_0 t)} - 1)t\eta(\gamma_0 t)dt, \quad (3.2.69)$$

say, and hence

$$E\sup_{|\gamma-\gamma_0|\leq\varepsilon}|\underline{v}_{i,n}(\gamma)|$$

$$\leq \frac{1}{2\pi\gamma_0^2}E\int|\underline{w}_{i,n}(t)|\sup_{|\gamma-\gamma_0|\leq\varepsilon}|\frac{\eta(\gamma t)}{\eta(\gamma_0 t)} - 1||t||\eta(\gamma_0 t)|dt$$

$$= \frac{1}{2\pi\gamma_0^2}\int E|\underline{w}_{i,n}(t)|\sup_{|\gamma-\gamma_0|\leq\varepsilon}|\frac{\eta(\gamma t)}{\eta(\gamma_0 t)} - 1||t||\eta(\gamma_0 t)|dt$$

$$\leq \frac{1}{2\pi\gamma_0^2}\int\sup_{|\gamma-\gamma_0|\leq\varepsilon}|\frac{\eta(\gamma t)}{\eta(\gamma_0 t)} - 1||t||\eta(\gamma_0 t)|dt.\sqrt{\frac{1}{n}\sum_{j=1}^n E\{(\partial/\partial\theta_i)g(\underline{x}_j,\theta_0)\}^2}, \quad (3.2.70)$$

because for all t,

$$E|\underline{w}_{i,n}(t)| \leq \sqrt{E\,\underline{w}_{i,n}(t)^2} \leq \sqrt{\frac{1}{n}\sum_{j=1}^n E\{(\partial/\partial\theta_i)g(\underline{x}_j,\theta_0)\}^2}. \quad (3.2.71)$$

Moreover, from the assumptions 3.1.3 and 3.1.4 and from theorem 2.2.15 it follows that:

$$\lim\frac{1}{n}\sum_{j=1}^n E\{(\partial/\partial\theta_i)g(\underline{x}_j,\theta_0)\}^2 = \int\{(\partial/\partial\theta_i)g(x,\theta_0)\}^2 dH(x) \quad (3.2.72)$$

and hence that (3.2.71) is uniformly bounded for $n \geq 1$. Thus putting

$K_i = \frac{1}{2\pi\gamma_0^2}\sup_n\sqrt{\frac{1}{n}\sum_{j=1}^n E\{(\partial/\partial\theta_i)g(\underline{x}_j,\theta_0)\}^2}$ it follows from (3.2.70) and Chebishev's inequality that for any $\delta > 0$ and any $\varepsilon > 0$

$$P\{|\underline{v}_{i,n}(\tilde{\underline{\gamma}}_n(\gamma_1))| > \delta\} = P\{|\underline{v}_{i,n}(\tilde{\underline{\gamma}}_n(\gamma_1))| > \delta \text{ and } |\tilde{\underline{\gamma}}_n(\gamma_1)-\gamma_0| \le \varepsilon\}$$

$$+ P\{|\underline{v}_{i,n}(\tilde{\underline{\gamma}}_n(\gamma_1))| > \delta \text{ and } |\tilde{\underline{\gamma}}_n(\gamma_1)-\gamma_0| > \varepsilon\}$$

$$\le P\{\sup_{|\gamma-\gamma_0|\le \varepsilon} |\underline{v}_{i,n}(\gamma)| > \delta\} + P\{|\tilde{\underline{\gamma}}_n(\gamma_1)-\gamma_0| > \varepsilon\}$$

$$\le \frac{E \sup_{|\gamma-\gamma_0|\le \varepsilon} |\underline{v}_{i,n}(\gamma)|}{\delta} + P\{|\tilde{\underline{\gamma}}_n(\gamma_1)-\gamma_0| > \varepsilon\}$$

$$\le K_i \int \sup_{|\gamma-\gamma_0|\le \varepsilon} \left|\frac{\eta(\gamma t)}{\eta(\gamma_0 t)} - 1\right| |t| |\eta(\gamma_0 t)| dt + P\{|\tilde{\underline{\gamma}}_n(\gamma_1)-\gamma_0| > \varepsilon\}. \quad (3.2.73)$$

If for example $\rho(u) = \frac{e^{-\frac{1}{2}u^2}}{\sqrt{2\pi}}$ then its characteristic function is $\eta(t)=e^{-\frac{1}{2}t^2}$ and thus

$$\sup_{|\gamma-\gamma_0|\le \varepsilon} \left|\frac{\eta(\gamma t)}{\eta(\gamma_0 t)} - 1\right| = \sup_{|\gamma-\gamma_0|\le \varepsilon} \left|e^{-\frac{1}{2}(\gamma^2-\gamma_0^2)t^2} - 1\right|$$

$$\le 1 - e^{-\varepsilon \sup_{\gamma\in\Gamma}|\gamma| t^2}, \quad (3.2.74)$$

so that

$$\lim_{\varepsilon\downarrow 0} \sup_{|\gamma-\gamma_0|\le \varepsilon} \left|\frac{\eta(\gamma t)}{\eta(\gamma_0 t)} - 1\right| = 0 . \quad (3.2.75)$$

From the bounded convergence theorem we then have

$$\lim_{\varepsilon\downarrow 0} \int \sup_{|\gamma-\gamma_0|\le \varepsilon} \left|\frac{\eta(\gamma t)}{\eta(\gamma_0 t)} - 1\right| |t| |\eta(\gamma_0 t)| dt = 0 , \quad (3.2.76)$$

while obviously (3.2.55) implies

$$\lim_{n\to\infty} P(|\tilde{\underline{\gamma}}_n(\gamma_1)-\gamma_0| > \varepsilon) = 0 . \quad (3.2.77)$$

From (3.2.73), (3.2.76) and (3.2.77) we now conclude that if

Assumption 3.2.4. $\rho(u)$ is chosen such that its characteristic function $\eta(t)$ satisfies

$$\int |t\eta(t)|dt < \infty \;, \; \sup_{t\in R} \sup_{|\gamma-\gamma_0|\leq \varepsilon} \left|\frac{\eta(\gamma t)}{\eta(\gamma_0 t)} - 1\right| < \infty$$

for some $\varepsilon > 0$ and

$$\lim_{\varepsilon\downarrow 0} \sup_{|\gamma-\gamma_0|\leq \varepsilon} \left|\frac{\eta(\gamma t)}{\eta(\gamma_0 t)} - 1\right| = 0 \quad \text{for any } t,$$

then for $i=1,2,\ldots,q$

$$\text{plim}\{\sqrt{n}(\partial/\partial\theta_i)\underline{R}_n(\theta_0|\tilde{\underline{\gamma}}_n(\gamma_1)) - \sqrt{n}(\partial/\partial\theta_i)\underline{R}_n(\theta_0|\gamma_0)\}=0, \tag{3.2.78}$$

which implies, using lemma 3.2.5, that

$$\sqrt{n}(\partial/\partial\theta')\underline{R}_n(\theta_0|\tilde{\underline{\gamma}}_n(\gamma_1)) \to N_q\left[0, \int \frac{1}{\gamma_0^4} \rho'(\frac{u}{\gamma_0})^2 f(u)du\, A(\theta_0)\right] \quad \text{in distr..} \tag{3.2.79}$$

Summarizing we now have:

Theorem 3.2.7. Let the conditions of theorem 3.2.2 be satisfied. Let Γ be a compact subinterval of $(0,\infty)$ and let γ_0 be a unique point in Γ satisfying $h(\gamma_0) = \inf_{\gamma\in\Gamma} h(\gamma)$. Then the statistic $\tilde{\underline{\gamma}}_n(\gamma_1)$ obtained from (3.2.54) is a strongly consistent estimator of γ_0: $\tilde{\underline{\gamma}}_n(\gamma_1)) \to \gamma_0$ a.s. for any $\gamma_1 > 0$. Moreover, the two-stage M-estimator $\tilde{\underline{\theta}}_n(\tilde{\gamma}_n(\gamma_1))$ is strongly consistent: $\tilde{\underline{\theta}}_n(\tilde{\underline{\gamma}}_n(\gamma_1)) \to \theta_0$ a.s. for any $\gamma_1 > 0$ and if in addition ρ satisfies assumption 3.2.4 then

$\sqrt{n}(\tilde{\underline{\theta}}_n(\tilde{\underline{\gamma}}_n(\gamma_1)) - \theta_0) \to N_q(0, h(\gamma_0)A(\theta_0)^{-1})$ in distr. and

$\underline{h}_n(\tilde{\underline{\gamma}}_n(\gamma_1), \tilde{\underline{\theta}}_n(\tilde{\underline{\gamma}}_n(\gamma_1)))\underline{A}_n(\tilde{\underline{\theta}}_n(\tilde{\underline{\gamma}}_n(\gamma_1)))^{-1}$ is a strongly consistent estimator of the asymptotic variance matrix involved.

3.2.7 Some weaker results

Finally, notice that the lemma's 3.2.3 and 3.2.5 and theorem 3.2.2 carry over when (3.2.15) only holds in probability, which is the case when only part a) of assumption 3.1.7 is satisfied. Moreover, it is easily seen that the "almost sure" and "strong" results in the theorem 3.2.6 and 3.2.7 then only hold in probality. Thus:

> Theorem 3.2.8. If only part a) of assumption 3.1.7 is satisfied then our previous results carry over, except that in the theorems 3.2.6 and 3.2.7 "a.s." and "strongly consistent" now becomes "in pr." and "weakly consistent", respectively.

3.3 Weighted nonlinear robust M-estimation

3.3.1 Introduction

Although our M-estimation theory requires less restrictive assumptions about the error distribution than least squares theory does, in both cases we need assumptions about the finiteness of moments of the regressors. But if we allow the moments of the error distribution to be infinite, why should we not also allow the moments of the regressors to be infinite?

While infiniteness of moments of the error distribution may seriously disturb the consistency or the speed of convergence of least squares estimators as we already have seen, infiniteness of moments of the regressors may just speed up the convergence. For example consider the simple regression model $\underline{y}_j=\theta_0\underline{x}_j+\underline{u}_j$, where the $\underline{u}_j$'s are random drawings from a distribution with zero expectation and finite variance σ^2 and the scalar regressors $\underline{x}_j$ are random drawings from a distribution with infinite second moment. Then the least-squares estimator $\hat{\underline{\theta}}_n$ satisfies (taking conditional expectations)

$$E(\hat{\underline{\theta}}_n-\theta_0)^2 = E\{E((\frac{\sum_{j=1}^n\underline{x}_j\underline{u}_j}{\sum_{j=1}^n\underline{x}_j^2})^2|\ \underline{x}_1,\ldots,\underline{x}_n\}=\frac{\sigma^2}{n}.E\{\frac{1}{\frac{1}{n}\sum_{j=1}^n\underline{x}_j^2}\}=o(\frac{1}{n}) \quad ,$$

because $\text{plim}(\frac{1}{\frac{1}{n}\sum_{j=1}^n\underline{x}_j^2}) = 0$ and hence from theorem 2.2.7, $E\{\frac{1}{\frac{1}{n}\sum_{j=1}^n\underline{x}_j^2}\} \to 0$. Under the usual condition that $\text{plim}\frac{1}{n}\sum_{j=1}^{n}\underline{x}_j^2$ is finite and non zero we only have $E(\hat{\underline{\theta}}_n - \theta_0)^2 = O(\frac{1}{n})$, which is obviously a less rapid rate of convergence.

If both the error distribution and the distributions of the regressors have infinite moments, the problem of consistency and limiting distribution of least squares estimators becomes rather messy, although Hannan and Kanter (1977) found for stationary autoregressions with errors (and hence regressors) having a stable distribution with characteristic exponent $\alpha<2$ that least squares estimates satisfy $n^{\frac{1}{\delta}}(\hat{\underline{\theta}}-\theta_0)\rightarrow 0$ a.s. for any $\delta>\alpha$. As a consequence, if for example the errors $\underline{u}_j$ of the autoregression $\underline{y}_j=\theta_0\underline{y}_{j-1} + \underline{u}_j$ are independent and identically Cauchy distributed (with zero mode) then the speed of adjustment of the least squares estimator

$$\hat{\underline{\theta}}_n=\theta_0 + \frac{\sum_{j=1}^{n}\underline{u}_j\underline{y}_{j-1}}{\sum_{j=1}^{n}\underline{y}_{j-1}^2}$$

is nearly twice that of the standard case, because now $n^{1-\delta}(\hat{\underline{\theta}}_n-\theta_0)\rightarrow 0$ in pr. for every $\delta>0$, instead of $(\sqrt{n})^{1-\delta}(\hat{\underline{\theta}}_n-\theta_0)\rightarrow 0$ in pr. for every $\delta>0$ in the standard case.
However, in this case the limiting distribution of $\hat{\underline{\theta}}_n-\theta_0$ is still unknown.

We may feel now that when the assumptions 3.1.7 and 3.2.2 fail to hold this is not disastrous for the consistency of M-estimates but on the contrary an advantage causing a more rapid convergence. But if so, this rapid convergence is paid for by vanishing knowledge about the limiting distribution.
In this section we shall therefore discuss a method for getting rid of the assumptions 3.1.7 and 3.2.2 without losing control of the asymptotic behaviour of the estimator involved, but also without being able to collect possible gains with respect to the speed of convergence.

The main idea is simple. Consider instead of (3.2.3) the objective function

$$\underline{R}_n^*(\theta|\gamma)= \frac{1}{n}\sum_{j=1}^{n} \frac{1}{\gamma}\, \rho\left(\frac{\underline{y}_j- g(\underline{x}_j,\theta)}{\gamma s(\underline{x}_j)}\right) \tag{3.3.1}$$

where

Assumption 3.3.1. $s(x)$ is any continuous function on R^p satisfying

$$\inf_{x\in R^p} s(x) > 0,\ \sup_{x\in R^p}\sup_{\theta\in \textcircled{H}}\left|\frac{(\partial/\partial\theta_i)g(x,\theta)}{s(x)}\right|< \infty$$

$$\text{and}\ \sup_{x\in R^p}\sup_{\theta\in \textcircled{H}}\left|\frac{(\partial/\partial\theta_{i_1})(\partial/\partial\theta_{i_2})g(x,\theta)}{s(x)}\right|< \infty \text{ for } i,i_1,i_2=1,2,\ldots,q.$$

(for example in the linear case $g(x,\theta)=x'\theta$ we may choose $s(x)=1+|x|$). Then

$$(\partial/\partial\theta_i)\underline{R}^*_n(\theta|\gamma)=-\frac{1}{n}\sum_{j=1}^n \frac{1}{\gamma^2}\rho'\left(\frac{\underline{y}_j - g(\underline{x}_j,\theta)}{\gamma s(\underline{x}_j)}\right).\frac{(\partial/\partial\theta_i)g(\underline{x}_j,\theta)}{s(\underline{x}_j)} \tag{3.3.2}$$

which is for every $\gamma>0$ and every $\theta\in\text{(H)}$ a mean of uniformly bounded random variables, and the same applies to

$$(\partial/\partial\theta_{i_1})(\partial/\partial\theta_{i_2})\underline{R}_n(\theta|\gamma)=$$

$$\frac{1}{n}\sum_{j=1}^n \frac{1}{\gamma^3}\rho''\left\{\frac{\underline{y}_j - g(\underline{x}_j,\theta)}{\gamma s(\underline{x}_j)}\right\}.\frac{(\partial/\partial\theta i_1)g(\underline{x}_j,\theta)}{s(\underline{x}_j)}\cdot\frac{(\partial/\partial\theta i_2)g(\underline{x}_j,\theta)}{s(\underline{x}_j)}$$

$$-\frac{1}{n}\sum_{j=1}^n \frac{1}{\gamma^2}\rho'\left\{\frac{\underline{y}_j - g(\underline{x}_j,\theta)}{\gamma s(\underline{x}_j)}\right\}.\frac{(\partial/\partial\theta_{i_1})(\partial/\partial\theta_{i_2})g(\underline{x}_j,\theta)}{s(\underline{x}_j)} \tag{3.3.3}$$

Thus the terms in (3.3.2) and (3.3.3) have finite moments of every order without using the assumptions 3.1.7 and 3.2.2.

3.3.2 Strong consistency and asymptotic normality

Since $s(x)$ is continuous it follows from the assumptions 3.1.1, 3.1.3 and 3.3.1 and theorem 2.3.3 that (3.2.6) carries over, where now

$$R^*(\theta|\gamma)=\iint\frac{1}{\gamma}\rho\left\{\frac{u + g(x,\theta_0)- g(x,\theta_0)}{\gamma s(x)}\right\}f(u)du\ dH(x) \tag{3.3.4}$$

Similar to (3.2.9) it follows from lemma 3.2.1 that for any $\gamma>0$ and any $x\in R^p$,

$$\int\frac{1}{\gamma}\rho\left(\frac{u}{\gamma s(x)}-z\right)f(u)du < \frac{1}{\gamma}\rho\left(\frac{u}{\gamma s(x)}\right)f(u)du \quad \text{if} \quad z \neq 0$$

and hence $\int\frac{1}{\gamma}\rho\left(\frac{u + g(x,\theta_0)- g(x,\theta)}{\gamma s(x)}\right)f(u)du < \int\frac{1}{\gamma}\rho\left(\frac{u}{\gamma s(x)}\right)f(u)du$ if

$\frac{g(x,\theta_0)- g(x,\theta)}{s(x)} \neq 0$. Thus if we assume

<u>Assumption 3.3.2.</u> $\int\left\{\frac{g(x,\theta)- g(x,\theta_0)}{s(x)}\right\}^2 dH(x)=0$ if and only if $\theta=\theta_0$

then similar to lemma 3.2.2, $R^*(\theta|\gamma)$ has for any $\gamma>0$ a unique maximum at $\theta=\theta_0$.

<u>Remark</u>: Note that from the mean value theorem,

$g(x,\theta)- g(x,\theta_0)=(\theta - \theta_0)'(\partial/\partial\theta')g(x,\theta_0 + \lambda(x)(\theta - \theta_0))$ for some $\lambda(x)\in[0,1]$.

If Θ is convex (assumption 3.1.6) then $\theta \in \Theta$ and $\theta_0 \in \Theta$ implies $\theta_0 + \lambda(x)(\theta - \theta_0) \in \Theta$, so that assumption 3.3.1 implies: $\sup_{x\in R^p} \sup_{\theta\in\Theta} \frac{|g(x,\theta)- g(x,\theta_0)|}{s(x)} < \infty$

and consequently $\int \{\frac{g(x,\theta)- g(x,\theta_0)}{s(x)}\}^2 dH(x) < \infty$ for all $\theta \in \Theta$.

Thus we conclude from lemma 3.1.3

<u>Theorem 3.2.1</u>. Choose ρ as in theorem 3.2.1. Under the assumptions 3.1.1, 3.2.1, 3.1.3, 3.3.1 and 3.3.2 we have $\tilde{\underline{\theta}}_n^*(\gamma) \to \theta_0$ a.s. for any $\gamma > 0$, where $\tilde{\underline{\theta}}_n^*(\gamma)$ is obtained from

$$\underline{R}_n^*(\tilde{\underline{\theta}}_n^*(\gamma)|\gamma) = \sup_{\theta\in\Theta} \underline{R}_n^*(\theta|\gamma) \quad , \quad \tilde{\underline{\theta}}_n^*(\gamma) \in \Theta \tag{3.3.5}$$

with $\underline{R}_n^*(\theta|\gamma)$ defined by (3.3.1)

We can be rather short about the limiting distribution of $\sqrt{n}(\underline{\theta}_n^*(\gamma) - \theta_0)$ because our argument in section 3.2.3 can be followed closely, except that the assumptions 3.1.7 and 3.2.2 are no longer needed. Thus let $\underline{\theta}_n^{**}(\gamma)$ be a random vector satisfying:

$$\sqrt{n}(\partial/\partial\theta)\underline{R}_n^*(\tilde{\underline{\theta}}_n^*(\gamma)|\gamma) = \sqrt{n}(\partial/\partial\theta)\underline{R}_n^*(\theta_0|\gamma) + \sqrt{n}(\tilde{\underline{\theta}}_n^*(\gamma)-\theta_0)(\partial/\partial\theta)(\partial/\partial\theta')\underline{R}_n^*(\underline{\theta}_n^{**}(\gamma)|\gamma). \tag{3.3.6}$$

Then under the assumptions 3.1.1 and 3.3.1:

$$\text{plim}(\partial/\partial\theta)(\partial/\partial\theta')\underline{R}_n^*(\underline{\theta}_n^{**}(\gamma)|\gamma) = B^*(\theta_0|\gamma) \text{ for any } \gamma > 0\, , \tag{3.3.7}$$

where

$B^*(\theta|\gamma) = (b_{i_1,i_2}^{**}(\theta|\gamma))$ with

$$b_{i_1,i_2}^{**}(\theta|\gamma) = \int\int \frac{1}{\gamma^3}\rho''\left(\frac{u+g(x,\theta_0)-g(x,\theta)}{\gamma s(x)}\right)f(u)du . \frac{(\partial/\partial\theta_{i_1})g(x,\theta)}{s(x)} . \frac{(\partial/\partial\theta_{i_2})g(x,\theta)}{s(x)} dH(x) \tag{3.3.8}$$

provided that ρ is chosen as in theorem 3.2.2. Note that under the assumptions of lemma 3.2.4, $\int \frac{1}{\gamma^3}\rho''(\frac{u}{\gamma . s})f(u)du < 0$ for any $\gamma > 0$ and any $s > 0$, while from the bounded convergence theorem, $\lim_{s\to\infty}\int \frac{1}{\gamma^3}\rho''(\frac{u}{\gamma . s})f(u)du = \frac{\rho''(0)}{\gamma^3}$.

Putting $\lambda_0 = \frac{1}{\inf s(x)}$ we thus see that

$$K_\gamma = \inf_{x\in R^p} -\int \frac{1}{\gamma^3}\rho''\left(\frac{u}{\gamma s(x)}\right)f(u)du \geq \inf_{\lambda\in[0,\lambda_0]} -\int \frac{1}{\gamma^3}\rho''\left(\lambda\frac{u}{\gamma}\right)f(u)du > 0 \quad \text{for any } \gamma > 0 ,$$

provided that $\rho''(0) < 0$. This implies that if we assume

Assumption 3.3.3. The matrix $C(\theta_0) = \left(\int \frac{(\partial/\partial\theta_{i_1})g(x,\theta_0)}{s(x)} \cdot \frac{(\partial/\partial\theta_{i_2})g(x,\theta_0)}{s(x)} dH(x)\right)$ is positive definite,

then for any non zero vector $\zeta \in R^q$, $-\zeta' B^*(\theta_0|\gamma)\zeta \geq K_\gamma \zeta' C(\theta_0)\zeta > 0$, hence $B^*(\theta_0|\gamma)$ is negative definite for any $\gamma > 0$.

Since $\int \rho'\left(\frac{u}{\gamma s(x)}\right)f(u)du = 0$ for any $x\in R^p$ and any $\gamma > 0$, it follows similar to lemma 3.2.5 that under assumption 3.1.1

$$\sqrt{n}(\partial/\partial\theta')\underline{R}_n^*(\theta_0|\gamma) \to N_q(0, A^*(\theta_0|\gamma)) \text{ in distr. for any } \gamma > 0 , \tag{3.3.9}$$

where now

$$A^*(\theta|\gamma) = (a^{**}_{i_1,i_2}(\theta|\gamma)) \text{ with } a^{**}_{i_1,i_2}(\theta|\gamma) = \iint \frac{1}{\gamma^4}\rho'\left(\frac{u+g(x,\theta_0)-g(x,\theta)}{\gamma s(x)}\right)^2 f(u)du.$$

$$\cdot \frac{(\partial/\partial\theta_{i_1})g(x,\theta)}{s(x)} \cdot \frac{(\partial/\partial\theta_{i_2})g(x,\theta)}{s(x)} dH(x) . \tag{3.3.10}$$

Moreover, also (3.2.13) carries over. Summarizing we now have:

Theorem 3.3.2. Choose ρ as in theorem 3.2.2 and such that $\rho''(0) < 0$. Moreover, let the assumptions of theorem 3.3.1 be satisfied. Under the additional assumptions 3.1.6, 3.2.3 and 3.3.3

$$\sqrt{n}(\tilde{\underline{\theta}}_n^*(\gamma) - \theta_0) \to N_q[0, B^*(\theta_0|\gamma)^{-1}A^*(\theta_0|\gamma)B^*(\theta_0|\gamma)^{-1}] \text{ in distr. for any } \gamma > 0.$$

3.3.3 A two-stage weighted robust M-estimator

Also now the theorem above is only a first step. In this subsection we shall derive a more efficient two-step weighted M-estimator. For this purpose we consider

$$\left.\begin{aligned}
&\underline{a}^{**}_{n,i_1,i_2}(\theta|\gamma)=\frac{1}{n}\sum_{j=1}^{n}\frac{1}{\gamma^4}\rho'\left(\frac{\underline{y}_j-g(\underline{x}_j,\theta)}{\gamma s(\underline{x}_j)}\right)^2\frac{(\partial/\partial\theta_{i_1})g(\underline{x}_j,\theta)}{s(\underline{x}_j)}\cdot\frac{(\partial/\partial\theta_{i_2})g(\underline{x}_j,\theta)}{s(\underline{x}_j)}\\
&\underline{b}^{**}_{n,i_1,i_2}(\theta|\gamma)=\frac{1}{n}\sum_{j=1}^{n}\frac{1}{\gamma^3}\rho''\left(\frac{\underline{y}_j-g(\underline{x}_j,\theta)}{\gamma s(\underline{x}_j)}\right)\frac{(\partial/\partial\theta_{i_1})g(\underline{x}_j,\theta)}{s(\underline{x}_j)}\cdot\frac{(\partial/\partial\theta_{i_2})g(\underline{x}_j,\theta)}{s(\underline{x}_j)}\\
&\underline{A}^{*}_{n}(\theta|\gamma)=(\underline{a}^{**}_{n,i_1,i_2}(\theta|\gamma))\ ;\ \underline{B}^{*}_{n}(\theta|\gamma)=(\underline{b}^{**}_{n,i_1,i_2}(\theta|\gamma))
\end{aligned}\right\}(3.3.11)$$

From the assumptions 3.1.1, 3.1.3 and 3.3.1 and theorem 2.3.3 we have:

$$\left.\begin{aligned}
&\sup_{(\theta,\gamma)\in \Theta\times\Gamma}|\underline{a}^{**}_{n,i_1,i_2}(\theta|\gamma)-a^{**}_{i_1,i_2}(\theta|\gamma)|\to 0\ \text{a.s.}\\
&\sup_{(\theta,\gamma)\in \Theta\times\Gamma}|\underline{b}^{**}_{n,i_1,i_2}(\theta|\gamma)-b^{**}_{i_1,i_2}(\theta|\gamma)|\to 0\ \text{a.s.}
\end{aligned}\right\}(3.3.12)$$

for any compact subinterval Γ of $(0,\infty)$. Since $a^{**}_{i_1,i_2}(\theta|\gamma)$ and $b^{**}_{i_1,i_2}(\theta|\gamma)$ are continuous on Θ x Γ and Θ x Γ is compact, it follows that they are uniformly continuous on Θ x Γ. Hence it follows from theorem 3.3.1 that:

$$\left.\begin{aligned}
&\sup_{\gamma\in\Gamma}|a^{**}_{i_1,i_2}(\tilde{\underline{\theta}}_n(\gamma_1)|\gamma)-a^{**}_{i_1,i_2}(\theta_0|\gamma)|\to 0\ \text{a.s. for any }\gamma_1>0\\
&\sup_{\gamma\in\Gamma}|b^{**}_{i_1,i_2}(\tilde{\underline{\theta}}_n(\gamma_1)|\gamma)-b^{**}_{i_1,i_2}(\theta_0|\gamma)|\to 0\ \text{a.s. for any }\gamma_1>0
\end{aligned}\right\}(3.3.13)$$

Obviously, from (3.3.12) and (3.3.13) we now have

$$\left.\begin{aligned}
&\sup_{\gamma\in\Gamma}|\underline{a}^{**}_{n,i_1,i_2}(\tilde{\underline{\theta}}^*_n(\gamma_1)|\gamma)-a^{**}_{i_1,i_2}(\theta_0|\gamma)|\to 0\ \text{a.s. for any }\gamma_1>0\\
&\sup_{\gamma\in\Gamma}|\underline{b}^{**}_{n,i_1,i_2}(\tilde{\underline{\theta}}^*_n(\gamma_1)|\gamma)-b^{**}_{i_1,i_2}(\theta_0|\gamma)|\to 0\ \text{a.s. for any }\gamma_1>0
\end{aligned}\right\}(3.3.14)$$

Consider the functions:

$$T(\gamma)=\text{trace}\left[B^*(\theta_0|\gamma)^{-1}A^*(\theta_0|\gamma)B^*(\theta_0|\gamma)^{-1}\right]\tag{3.3.15}$$

$$\underline{T}_n(\gamma|\gamma_1)=\text{trace}\left[\underline{B}^*_n(\tilde{\underline{\theta}}^*_n(\gamma_1)|\gamma)^{-1}\underline{A}^*_n(\tilde{\underline{\theta}}^*_n(\gamma_1)|\gamma)\underline{B}^*_n(\tilde{\underline{\theta}}^*_n(\gamma_1)|\gamma)^{-1}\right]\tag{3.3.16}$$

Since the elements of the inverse of a non singular matrix are continuous functions of the elements of the latter matrix, it follows that $T(\gamma)$ is a continuous function on any set Γ such that $B^*(\theta_0|\gamma)$ is non singular for all $\gamma\in\Gamma$, because the elements of $B^*(\theta_0|\gamma)$ are continuous functions of $\gamma > 0$. Suppose

Assumption 3.3.4. There is a unique point γ_0 in a compact subinterval Γ of $(0,\infty)$ such that $T(\gamma_0) = \inf_{\gamma\in\Gamma} T(\gamma)$,

and let the r.v. $\tilde{\underline{\gamma}}_n^*(\gamma_1)$ in Γ be a solution of

$$\underline{T}_n(\tilde{\underline{\gamma}}_n^*(\gamma_1)|\gamma_1) = \inf_{\gamma\in\Gamma} \underline{T}_n(\gamma|\gamma_1). \tag{3.3.17}$$

We shall prove now that:

Theorem 3.3.3. Under the assumptions of theorem 3.3.2 and the additional assumption 3.3.4,

$$\tilde{\underline{\gamma}}_n^*(\gamma_1) \to \gamma_0 \quad \text{a.s. for any } \gamma_1 > 0 \tag{3.3.18}$$

and

$$\underline{B}_n^*(\tilde{\underline{\theta}}_n^*(\gamma_1)|\tilde{\underline{\gamma}}_n^*(\gamma_1))^{-1}\,\underline{A}_n^*(\tilde{\underline{\theta}}_n^*(\gamma_1)|\tilde{\underline{\gamma}}_n^*(\gamma_1))\,\underline{B}_n^*(\tilde{\underline{\theta}}_n^*(\gamma_1)|\tilde{\underline{\gamma}}_n^*(\gamma_1))^{-1}$$

$$\to B^*(\theta_0|\gamma_0)^{-1}A^*(\theta_0|\gamma_0)\;B^*(\theta_0|\gamma_0)^{-1} \text{ a.s.} \tag{3.3.19}$$

for any $\gamma_1 > 0$.

Proof: Part (3.3.18) follows from (3.3.14), assumption 3.3.4, lemma 3.1.3 and lemma 3.3.1 below. Then by (3.3.14), (3.3.18) and the continuity of $a_{i_1,i_2}(\theta_0|\gamma)$ and $b_{i_1,i_2}(\theta_0|\gamma)$ on Γ it follows from theorem 2.3.2

$$\left.\begin{array}{l} \underline{A}_n^*(\tilde{\underline{\theta}}_n^*(\gamma_1)|\tilde{\underline{\gamma}}_n^*(\gamma_1)) \to A^*(\theta_0|\gamma_0) \text{ a.s. for any } \gamma_1 > 0 \\ \underline{B}_n^*(\tilde{\underline{\theta}}_n^*(\gamma_1)|\tilde{\underline{\gamma}}_n^*(\gamma_1)) \to B^*(\theta_0|\gamma_0) \text{ a.s. for any } \gamma_1 > 0 \end{array}\right\} \tag{3.3.20}$$

Since the elements of $B^{-1}AB^{-1}$ are continuous functions of those of A and B, provided that B is non singular, (3.3.19) follows from (3.3.20) and theorem 2.2.5.
□

<u>Lemma 3.3.1.</u> Let $(\underline{a}_{i,n}(\theta))$, $i=1,2,\ldots,k$ be random functions on a compact subset Ⓗ of R^m. Let $a_1(\theta),\ldots,a_k(\theta)$ be continuous nonrandom functions on Ⓗ such that

$$\left.\begin{array}{l} \text{a) } \underline{a}_{i,n}(\theta)\rightarrow a_i(\theta) \text{ a.s. pseudo-uniformly on Ⓗ for } i=1,2,\ldots,k \\ \text{or} \\ \text{b) } \underline{a}_{i,n}(\theta)\rightarrow a_i(\theta) \text{ in pr. pseudo-uniformly on Ⓗ for } i=1,2,\ldots k \end{array}\right\} \quad (3.3.21)$$

respectively. Let $f(x_1,\ldots,x_k)$ be a continuous function on an open subset X of R^k, where X is such that $(a_1(\theta),\ldots,a_k(\theta))\in X$ for all $\theta\in$Ⓗ

Then:

a) $f(\underline{a}_{1,n}(\theta),\ldots,\underline{a}_{k,n}(\theta))\rightarrow f(a_1(\theta),\ldots,a_k(\theta))$ a.s. pseudo-uniformly on Ⓗ

or:

b) $f(\underline{a}_{1,n}(\theta),\ldots,\underline{a}_{k,n}(\theta))\rightarrow f(a_1(\theta),\ldots,a_k(\theta))$ in pr. pseudo-uniformly on Ⓗ,

respectively.

<u>Proof</u>: a) Since Ⓗ is a compact subset of R^k and the $a_i(\theta)$'s are continuous functions on Ⓗ, it follows from Rudin (1976, theorem 4.15) that the range of $(a_1(\theta),\ldots,a_k(\theta))$ is closed and bounded, and hence compact. This implies that there is a compact subset C of X such that $(a_1(\theta),\ldots,a_k(\theta))\in C$ for all $\theta\in$Ⓗ. Since C is closed and X is open, there is an $\varepsilon_0>0$ such that:

$$\{x\in R^k: |x-c|<\varepsilon_0\}\in X \quad \text{for all } c\in C.$$

Hence, putting

$$C_1=\bigcup_{c\in C}\{x\in R^k: |x-c|<\varepsilon_0\}$$

we have

$$C\subset C_1\subset X\ .$$

Next, let

$$C_2=\bigcup_{c\in C}\{x\in R^k: |x-c|<\tfrac{1}{2}\varepsilon_0\} \quad (3.3.22)$$

and let $\overline{C}_2$ be the closure of C_2. It is not hard to prove that $\overline{C}_2\subset C_1$. Since $\overline{C}_2$ is closed and C_1 is bounded (because C is bounded), $\overline{C}_2$ is closed and bounded and hence compact.

Now let $\{\Omega,\mathcal{F},p\}$ be the probability space involved, and let N_* be the null set on which (3.3.21/a) fails to hold. Choose ε arbitrarily in $(0,\frac{1}{2}\varepsilon_0)$. From (3.3.21a) it follows that for every $\omega\in\Omega\setminus N_*$ there is a number $n_0(\omega,\varepsilon)$ such that for all $\theta\in$ Ⓗ

$$|(a_{1,n}(\omega,\theta),\ldots,a_{k,n}(\omega,\theta))-(a_1(\theta),\ldots,a_k(\theta))|\leq\varepsilon \text{ if } n\geq n_0(\omega,\varepsilon). \qquad (3.3.23)$$

But since

$$(a_1(\theta),\ldots,a_k(\theta))\in C \qquad (\subset\overline{C}_2) \qquad (3.3.24)$$

and since $\varepsilon<\frac{1}{2}\varepsilon_0$, (3.3.22) and (3.3.23) imply:

$$(a_{1,n}(\omega,\theta),\ldots,a_{k,n}(\omega,\theta))\in C_2\subset\overline{C}_2 \text{ for all } \theta\in \text{Ⓗ if } n\geq n_0(\omega,\varepsilon). \qquad (3.3.25)$$

Moreover, since f is continuous on the compact set $\overline{C}_2$, f is uniformly continuous on $\overline{C}_2$.

Therefore it follows from (3.3.23), (3.3.24) and (3.3.25) that for any $\delta>0$ we can choose ε so small that for all $\omega\in\Omega\setminus N_*$ and all $\theta\in$ Ⓗ ,

$$|f(a_{1,n}(\omega,\theta),\ldots,a_{k,n}(\omega,\theta))-f(a_1(\theta),\ldots,a_k(\theta))|<\delta \text{ if } n\geq n_0(\omega,\varepsilon)$$

This proves part a). Part b) follows directly from a), using theorem 2.2.4.

□

Our next step is to show that an analogue of theorem 3.2.7 holds. First, similar to (3.2.56), (3.2.62) and (3.2.63) it can be shown that under the assumptions of theorem 3.3.3

$$\underline{\tilde{\theta}}_n^*(\underline{\tilde{\gamma}}_n^*(\gamma_1))\to\theta_0 \text{ a.s. for any } \gamma_1>0, \qquad (3.3.26)$$

$$(\partial/\partial\theta)(\partial/\partial\theta')\underline{R}_n^*(\underline{\theta}_n^{**}(\underline{\tilde{\gamma}}_n^*(\tilde{\gamma}_1))|\underline{\tilde{\gamma}}_n^*(\gamma_1))\to B(\theta_0|\gamma_0) \text{ a.s. for any } \gamma_1>0 \qquad (3.3.27)$$

and

$$\sqrt{n}(\partial/\partial\theta)\underline{R}_n^*(\underline{\tilde{\theta}}_n^*(\underline{\tilde{\gamma}}_n(\gamma_1))|\underline{\tilde{\gamma}}_n^*(\gamma_1))\to 0' \text{ a.s. for any } \gamma_1>0\,. \qquad (3.3.28)$$

In the present case, (3.2.69) now becomes

$$\underline{v}_{i,n}^*(\gamma)=\sqrt{n}(\partial/\partial\theta_i)\underline{R}_n^*(\theta_0|\gamma)-\sqrt{n}(\partial/\partial\theta_i)\underline{R}_n^*(\theta_0|\gamma_0)$$

$$=-\frac{1}{2\pi\gamma_0^2}\int\{\frac{1}{\sqrt{n}}\sum_{j=1}^n\sin(t\frac{\underline{u}_j}{s(\underline{x}_j)})\frac{(\partial/\partial\theta_i)g(\underline{x}_j,\theta_0)}{s(\underline{x}_j)}\}t\{\eta(\gamma t)-\eta(\gamma_0 t)\}dt$$

$$=-\frac{1}{2\pi\gamma_0^2}\int\underline{w}_{i,n}^*(t)(\frac{\eta(\gamma t)}{\eta(\gamma_0 t)}-1)t\eta(\gamma_0 t)dt\,, \qquad (3.3.29)$$

say, while (3.2.71) becomes

$$E|\underline{w}^*_{i,n}(t)| \leq \sqrt{\frac{1}{n}\sum_{j=1}^{n} E\{\frac{(\partial/\partial\theta_i)g(\underline{x}_j,\theta_0)}{s(\underline{x}_j)}\}^2} \leq \sup_{x\in R^p}\frac{|(\partial/\partial\theta_i)g(x,\theta_0)|}{s(x)} < \infty . \tag{3.3.30}$$

Thus similar to (3.2.78) we have under assumption 3.2.4

$$\text{plim } \underline{v}^*_{i,n}(\tilde{\underline{\gamma}}^*_n(\gamma_1)) = 0 \quad \text{for any } \gamma_1 > 0 \tag{3.3.31}$$

and hence we conclude from (3.3.26), (3.3.27), (3.3.28) and (3.3.31):

Theorem 3.3.4. Under the assumptions of theorem 3.3.3 we have $\tilde{\underline{\theta}}^*_n(\tilde{\underline{\gamma}}^*_n(\gamma_1)) \to \theta_0$ a.s. for any $\gamma_1 > 0$.

If in addition also assumption 3.2.4 is satisfied then $\sqrt{n}(\tilde{\underline{\theta}}^*_n(\tilde{\underline{\gamma}}^*_n(\gamma_1)) - \theta_0) \to N_q[0, B^*(\theta_0|\gamma_0)^{-1}A^*(\theta_0|\gamma_0)B^*(\theta_0|\gamma_0)^{-1}]$ in distr.

3.4 Miscellaneous notes on robust M-estimation

3.4.1 Uniform consistency

In theorem 3.2.1 we concluded from (3.2.6) and lemma 3.2.2 that $\tilde{\underline{\theta}}_n(\gamma) \to \theta_0$ a.s. for every $\gamma > 0$. But the same conclusion can be drawn when instead of (3.2.6) only :

$$\sup_{\theta\in \textcircled{H}} |\underline{R}_n(\theta|\gamma) \to R(\theta|\gamma)| \to 0 \qquad \text{a.s. pointwise in } \gamma>0.$$

So we feel now that the conclusion $\tilde{\underline{\theta}}_n(\gamma) \to \theta_0$ a.s. for every $\gamma > 0$ may be too weak, and that possibly:

$$\tilde{\underline{\theta}}_n(\gamma) \to \theta_0 \text{ a.s. pseudo-uniformly on } \Gamma \tag{3.4.1}$$

for any compact interval Γ of $(0,\infty)$. Similarly we may feel that the conclusion of theorem 3.3.1 should be :

$$\tilde{\underline{\theta}}^*_n(\gamma) \to \theta_0 \text{ a.s. pseudo-uniformly on } \Gamma. \tag{3.4.2}$$

The following generalization of lemma 3.1.3 shows that this feeling is indeed right.

Lemma 3.4.1. Let $(\underline{Q}_n(\theta,\gamma))$ be a sequence of random functions on $\Theta \times \Gamma$, where Θ is a compact subset of R^k and Γ is a compact subset of R^m. Suppose $\underline{Q}_n(\theta,\gamma)\to Q(\theta,\gamma)$ a.s. pseudo-uniformly on $\Theta \times \Gamma$, where $Q(\theta,\gamma)$ is a continuous non-random function on $\Theta \times \Gamma$. If there is a point θ_0 in Θ such that $Q(\theta_0,\gamma) = \inf_{\theta\in\Theta} Q(\theta,\gamma)$ for every $\gamma\in\Gamma$ and if this point θ_0 is unique for every $\gamma\in\Gamma$ (thus for every $\gamma\in\Gamma$, $Q(\theta_0,\gamma) = Q(\theta,\gamma)$ implies $\theta=\theta_0$) then for any random vector function $\hat{\underline{\theta}}_n(\gamma)$ on Γ satisfying $\underline{Q}_n(\hat{\underline{\theta}}_n(\gamma),\gamma) = \inf_{\theta\in\Theta} \underline{Q}_n(\theta,\gamma)$, $\hat{\underline{\theta}}_n(\gamma)\in\Theta$ a.s., $\gamma\in\Gamma$, we have

$\hat{\underline{\theta}}_n(\gamma)\to\theta_0$ a.s. pseudo-uniformly on Γ.

Proof: First we show that:

$$Q(\hat{\underline{\theta}}_n(\gamma),\gamma)\to Q(\theta_0,\gamma) \text{ a.s. pseudo-uniformly on } \Gamma. \tag{3.4.3}$$

Let $\{\Omega,\mathfrak{F},P\}$ be the probability space. We have for $\omega\in\Omega\setminus N$, where N is some null set,

$$Q_n(\hat{\theta}_n(\gamma,\omega),\gamma,\omega) - Q(\theta_0,\gamma) =$$

$$= Q_n(\hat{\theta}_n(\gamma,\omega),\gamma,\omega) - Q(\hat{\theta}_n(\gamma,\omega),\gamma) + Q(\hat{\theta}_n(\gamma,\omega),\gamma) - Q(\theta_0,\gamma)$$

$$\begin{cases} \geq Q_n(\hat{\theta}_n(\gamma,\omega),\gamma,\omega) - Q(\hat{\theta}_n(\gamma,\omega),\gamma) \\ \leq Q_n(\theta_0,\gamma,\omega) - Q(\theta_0,\gamma) \end{cases} \tag{3.4.4}$$

hence

$$\sup_{\gamma\in\Gamma}|Q_n(\hat{\theta}_n(\gamma,\omega),\gamma,\omega) - Q(\theta_0,\gamma)| \leq$$

$$\leq \sup_{(\theta,\gamma)\in\Theta\times\Gamma} |Q_n(\theta,\gamma,\omega) - Q(\theta,\omega)| \tag{3.4.5}$$

and

$$\sup_{\gamma\in\Gamma}|Q(\hat{\theta}_n(\gamma,\omega),\gamma) - Q(\theta_0,\gamma)| \leq$$

$$\leq \sup_{(\theta,\gamma)\in\Theta\times\Gamma} |Q_n(\theta,\gamma,\omega) - Q(\theta,\gamma)| + \sup_{\gamma\in\Gamma}|Q_n(\hat{\theta}_n(\gamma,\omega),\gamma,\omega) - Q(\theta_0,\gamma)| \tag{3.4.6}$$

Since $\underline{Q}_n(\theta,\gamma)\to Q(\theta,\gamma)$ a.s. pseudo-uniformly on $\Theta \times \Gamma$, (3.4.3) follows now from (3.4.5) and (3.4.6).

From lemma 3.1.3 it follows that $\hat{\underline{\theta}}_n(\gamma)\to\theta_0$ a.s. for every $\gamma\in\Gamma$, or with other words: for every $\gamma\in\Gamma$ there is a null set N_γ with respect to the probability space $\{\Omega,\mathcal{F},P\}$ such that for every $\varepsilon > 0$ and every $\omega\in\Omega\setminus N_\gamma$ we can find a number $n_0(\varepsilon,\omega,\gamma)$ so that:

$$|\hat{\theta}_n(\omega,\gamma) - \theta_0| \leq \varepsilon \quad \text{for all } n \geq n_0(\varepsilon,\omega,\gamma),$$

and without loss of generality we may assume that $n_0(\varepsilon,\omega,\gamma)$ is the minimal number with this property. However, using (3.4.3) instead of (3.1.5) in the proof of lemma 3.1.3, we see that we may choose $N_\gamma=N$ for all $\gamma\in\Gamma$, where N is the null set on which (3.4.3) fails to hold.

Now suppose that $\sup_{\gamma\in\Gamma} n_0(\varepsilon,\omega_0,\gamma)=\infty$ for some $\varepsilon > 0$ and some $\omega_0\in\Omega\setminus N$. Then there is a sequence (γ_j) of points in Γ such that $n_0(\varepsilon,\omega_0,\gamma_j)\to\infty$ as $j\to\infty$. Moreover, then we can also find a subsequence (n_j) such that:

$$n_j\to\infty \text{ as } j\to\infty, \; |\hat{\theta}_{n_j}(\omega_0,\gamma_j)-\theta_0|>\varepsilon \quad \text{for all } j \; . \tag{3.4.7}$$

But (3.4.3) implies that:

$$|Q(\hat{\theta}_{n_j}(\omega_0,\gamma_j),\gamma_j)- Q(\theta_0,\gamma_j)|\leq \sup_{\gamma\in\Gamma}|Q(\hat{\theta}_{n_j}(\omega_0,\gamma),\gamma)- Q(\theta_0,\gamma)| \to 0 . \tag{3.4.8}$$

Moreover, $Q(\theta,\gamma)$ is continuous on the compact set $\Theta \times \Gamma$, and hence uniformly continuous on $\Theta \times \Gamma$. So (3.4.8) contradicts (3.4.7), hence:

$$\sup_{\gamma\in\Gamma} n_0(\varepsilon,\omega,\gamma)=n_*(\varepsilon,\omega)< \infty \quad \text{for every } \varepsilon>0 \text{ and every } \omega\in\Omega\setminus N.$$

Consequently, for every $\varepsilon > 0$ and every $\omega\in\Omega\setminus N$ we have:

$$|\hat{\theta}_n(\omega,\gamma)- \theta_0|\leq \varepsilon \text{ for all } n\geq n_*(\varepsilon,\omega),$$

which by definition 2.3.2 proves the lemma. □

We now conclude that under the conditions of theorem 3.2.1 we have $\tilde{\underline{\theta}}_n(\gamma)\to\theta_0$ a.s. pseudo-uniformly on a compact interval Γ, and similarly we conclude that under the conditions of theorem 3.3.3, $\tilde{\underline{\theta}}_n^*(\gamma)\to\theta_0$ a.s. pseudo-uniformly on Γ. But both the random functions $\underline{R}_n(\theta,\gamma)$ and $\underline{R}_n^*(\theta,\gamma)$ are continuous functions of θ,γ, and the observations $(\underline{y}_1,\underline{x}_1),\dots,(\underline{y}_n,\underline{x}_n)$, as is easily observed from (3.2.3) and (3.3.1), respectively. From lemma 3.1.2 it follows therefore that if $\tilde{\underline{\theta}}_n(\gamma)$ and $\tilde{\underline{\theta}}_n^*(\gamma)$ are unique then they are continuous vector functions of γ and the observations, hence by theorem 2.3.1, $\sup_{\gamma\in\Gamma}|\tilde{\underline{\theta}}_n(\gamma)-\theta_0|$ and $\sup_{\gamma\in\Gamma}|\tilde{\underline{\theta}}_n^*(\gamma)-\theta_0|$ are random variables. So we have proved by now:

> <u>Theorem 3.4.1.</u> Under the conditions of theorem 3.2.1 we have $\tilde{\underline{\theta}}_n(\gamma)\to\theta_0$ a.s. uniformly on Γ and under the conditions of theorem 3.3.1 we have $\tilde{\underline{\theta}}_n^*(\gamma)\to\theta_0$ a.s. uniformly on Γ, for any compact interval Γ of $(0,\infty)$, provided that $\tilde{\underline{\theta}}_n(\gamma)$ and $\tilde{\underline{\theta}}_n^*(\gamma)$ are unique. Otherwise we have only pseudo-uniform convergence.

3.4.2 The symmetric unimodality assumption

In the sections 3.2 and 3.3 the assumption that the error distribution is symmetric unimodal plays a key role, because it assures that the a.s. limits of the objective function $\underline{R}_n(\theta|\gamma)$ and $\underline{R}_n^*(\theta|\gamma)$ have a unique supremum at $\theta=\theta_0$, which "identifies" the parameter vector. However, as far as the results of section 3.2 are concerned this assumption may be weakened by the following:

> Assumption 3.4.1. Given a continuous bounded density $\rho(.)$, the error distribution $F(u)$ is such that the convolution
>
> $\phi(z|\gamma)=\int \frac{1}{\gamma}\rho(\frac{u-z}{\gamma})dF(u)$
>
> has a unique maximal mode at $\delta(\gamma)$, say, for every γ in a subset Γ of $(0,\infty)$.

Thus there may be more than one mode, but for every $\gamma\in\Gamma$ only one gives the highest value of $\phi(z|\gamma)$. Furthermore, suppose that

> Assumption 3.4.2. One of the components of θ_0, say the first one, is a constant term.

Under the assumptions 3.1.1, 3.1.3, 3.4.1 and 3.4.2 we now have

$$R(\theta_0+ \delta(\gamma)e_1|\gamma)= \sup_{\theta\in(H)} R(\theta|\gamma) , \qquad (3.4.9)$$

provided that:

> Assumption 3.4.3. $\theta_0+ \delta(\gamma)e_1 \in (H)$ for $\gamma\in\Gamma$,

where e_1 is a vector with 1 as first component and zero's at the other places. For it follows from (3.2.7) and the three assumptions above that

$$g(x,\theta_0+ \delta(\gamma)e_1)- g(x,\theta_0)= \delta(\gamma) \quad \text{everywhere in } x. \qquad (3.4.10)$$

Putting $\theta_0(\gamma)=\theta_0 + \delta(\gamma)e_1$ it can now be shown along the same lines as for theorem 3.2.1 that:

> Theorem 3.4.2. Under the assumptions 3.1.1, 3.1.3, 3.1.5, 3.4.1 through 3.4.3 we have $\underline{\tilde{\theta}}_n(\gamma) \to \theta_0(\gamma)$ a.s. for every $\gamma\in\Gamma$.

Thus the parameters of model(3.1.1), except the constant term, can be estimated strongly consistent by our M-estimation method even if the error distribution is not symmetric unimodal. It is not hard to show that also the theorems 3.2.2, 3.2.7 and 3.2.8 carry over for the case under review, provided that θ_0 is replaced with $\theta_0(\gamma)$ and the function $h(\gamma)$ is redefined by

$$h(\gamma) = \frac{h_1(\gamma)}{h_2(\gamma)^2} = \frac{\gamma^2 \int \rho'(\frac{u-\delta(\gamma)}{\gamma})^2 dF(u)}{\{\int \rho''(\frac{u-\delta(\gamma)}{\gamma}) dF(u)\}^2} \quad . \tag{3.4.11}$$

Even the theorems 3.2.3 and 3.2.4 carry over, provided that in addition:

$$\frac{\delta(\gamma)}{\gamma} \to 0 \quad \text{as} \quad \gamma \to \infty \quad . \tag{3.4.12}$$

For the case considered in section 3.3 the symmetric unimodality assumption of the error distribution can hardly be weakened. It appears that this assumption can only be replaced by assumption 3.4.1 if the maximal mode $\delta(\gamma)$ is independent of γ, which will be rare.

3.4.3 The function ρ

Throughout the sections 3.2 and 3.3 it is assumed that the function ρ is everywhere (twice) differentiable. This excludes a number of functions that are popular in the statistical literature on robust M-estimation, such as $\rho(u)=-|u|^p$, $p>0$ (p-norm estimates) and:

$$\rho(u) = \begin{cases} a^2 - u^2 & \text{if } |u| \le a \\ 0 & \text{if } |u| > a \end{cases} \quad , \quad a > 0 \tag{3.4.13}$$

(truncated least squares). But as long as the function ρ is chosen to be continuous and such that $\psi(z) = \int \rho(u-z) dF(u)$ has a unique maximum at $z=0$, probably also then the consistency of the nonlinear M-estimator involved can be proved along the same lines as we did. Moreover, if $\rho'(u)$ and $\rho''(u)$ are continuous in some neighbourhood of zero, the Taylor expansion approach can also now be used for establishing asymptotic normality. Probably we then need a generalization of theorem 2.2.16 and hence theorem 2.2.10 for discontinuous functions ϕ. However, if the number of discontinuities of ϕ is finite and the jumps involved are also finite, the latter seems not too hard to do.

3.4.4 How to decide to apply robust M-estimation

From theorem 3.2.4 we see that our robust M-estimation method is more efficient than least squares estimation if condition (3.2.30) holds. This suggest that the "residual kurtosis":

$$\frac{1}{n}\sum_{j=1}^{n} \hat{u}_j^4 \cdot \{\frac{1}{n}\sum_{j=1}^{n} \hat{u}_j^2\}^{-2} - 3 \quad ,$$

where the $\hat{u}_j$'s are the residuals, may serve as a criterion for this condition. However, in order to ensure that $\text{plim} \frac{1}{n}\sum_{j=1}^{n} \hat{u}_j^4 = \text{plim} \frac{1}{n}\sum_{j=1}^{n} u_j^4$ we need additional assumptions, even if $\int u^4 dF(u) du < \infty$. For example, let $g(\underline{x}_j, \theta) = \underline{x}_j \cdot \theta$, where both θ and $\underline{x}_j$ are scalars, and let $\hat{\theta}_n$ be the least squares estimator. Then $\hat{\underline{u}}_j = u_j + \underline{x}_j(\theta_0 - \hat{\theta}_n)$, hence

$$\frac{1}{n}\sum_{j=1}^{n}\hat{u}_j^4 - \frac{1}{n}\sum_{j=1}^{n}u_j^4 =$$
$$(\theta_0 - \hat{\theta}_n)^4 \frac{1}{n}\sum_{j=1}^{n}x_j^4 + 4(\theta_0 - \hat{\theta}_n)^3 \frac{1}{n}\sum_{j=1}^{n}x_j^3 u_j +$$
$$+6(\theta_0 - \hat{\theta}_n)^2 \frac{1}{n}\sum_{j=1}^{n}x_j^2 u_j^2 + 4(\theta_0 - \hat{\theta}_n)\frac{1}{n}\sum_{j=1}^{n}x_j u_j^3 \;.$$

Since under the conditions of theorem 3.1.2, $\sqrt{n}(\hat{\theta}_n - \theta_0)$ converges in distribution to the normal, it follows that $\text{plim}\{\frac{1}{n}\sum_{j=1}^{n}\hat{u}_j^4 - \frac{1}{n}\sum_{j=1}^{n}u_j^4\}=0$ if $\text{plim}\ \frac{1}{n^3}\sum_{j=1}^{n}x_j^4=0$, $\text{plim}(\frac{1}{\sqrt{n}})^3 \frac{1}{n}\sum_{j=1}^{n}x_j^3 u_j=0$, $\text{plim}\frac{1}{n^2}\sum_{j=1}^{n}x_j^2 u_j^2=0$ and $\text{plim}\ \frac{1}{n\sqrt{n}}\sum_{j=1}^{n}x_j u_j^3=0$. Of course, the latter conditions seems to be very weak, but they are not automatically implied by the conditions of theorem 3.1.2, even if $\int u^4 f(u)du<\infty$. However, the approach of section 3.3,where we used a weight function s(x),can also now be applied to solve this problem.

Consider model 3.1.1 and let s(x) be a continuous function on R^p such that

$$M=\inf_{x\in R^p} s(x)>0,\ K=\sup_{x\in R^p}\{\frac{\sup_{\theta\in\Theta}|(\partial/\partial\theta)g(x,\theta)|}{s(x)}\}<\infty \;. \tag{3.4.14}$$

Moreover, let $\hat{\theta}_n$ be a consistent estimator of θ_0. Then by the mean value theorem

$$\left|\frac{\hat{u}_j - u_j}{s(x_j)}\right| = \left|\frac{g(x_j,\theta_0) - g(x_j,\hat{\theta}_n)}{s(x_j)}\right| \le |\hat{\theta}_n - \theta_0|K$$

if $\hat{\theta}_n \in \Theta$, hence:

$$\left|\frac{1}{n}\sum_{j=1}^{n}\left(\frac{\hat{u}_j}{s(x_j)}\right)^4 - \frac{1}{n}\sum_{j=1}^{n}\left(\frac{u_j}{s(x_j)}\right)^4\right| \le$$
$$\le |\hat{\theta}_n - \theta_0|^4 K^4 + 4|\hat{\theta}_n - \theta_0|^3 \frac{K^3}{M} \frac{1}{n}\sum_{j=1}^{n}|u_j| +$$
$$+6|\hat{\theta}_n - \theta_0|^2 \frac{K^2}{M^2} \frac{1}{n}\sum_{j=1}^{n}u_j^2 + 4|\hat{\theta}_n - \theta_0|\frac{K}{M^3}\frac{1}{n}\sum_{j=1}^{n}|u_j|^3 \tag{3.4.15}$$

if $\hat{\theta}_n \in \Theta$, and consequently,

$$\text{plim}\{\frac{1}{n}\sum_{j=1}^{n}\left(\frac{\hat{u}_j}{s(x_j)}\right)^4 - \frac{1}{n}\sum_{j=1}^{n}\left(\frac{u_j}{s(x_j)}\right)^4\}=0 \;, \tag{3.4.16}$$

provided that $\int u^4 dF(u)<\infty$. Next we show

$$\text{plim}\ \frac{1}{n}\sum_{j=1}^{n}\left(\frac{u_j}{s(x_j)}\right)^4 = \int u^4 dF(u)\int\left(\frac{1}{s(x)}\right)^4 dH(x) \;. \tag{3.4.17}$$

Let $v_{n,j}=\begin{cases} u_j & \text{if } |u_j|\le a_n \\ 0 & \text{if } |u_j|\ge a_n \end{cases}$, where (a_n) is a sequence of positive numbers satisfying

$$\lim_{n\to\infty} a_n = \infty,\quad \lim_{n\to\infty}\frac{a_n^8}{n}=0 \;.$$

Then

$$E\left|\frac{1}{n}\sum_{j=1}^{n}\left\{\left(\frac{u_j}{s(\underline{x}_j)}\right)^4 - E\left(\frac{u_j}{s(\underline{x}_j)}\right)^4\right\} - \frac{1}{n}\sum_{j=1}^{n}\left\{\left(\frac{v_{n,j}}{s(\underline{x}_j)}\right)^4 - E\left(\frac{v_{n,j}}{s(\underline{x}_j)}\right)^4\right\}\right| \leq$$

$$\leq 2\,\frac{1}{n}\sum_{j=1}^{n}\frac{E|u_j^4 - v_{n,j}^4|}{M^4} = \frac{2}{M^4}\left\{\int_{a_n}^{\infty} u^4 dF(u) + \int_{-\infty}^{-a_n} u^4 dF(u)\right\} \to 0 \text{ as } n\to\infty \quad (3.4.18)$$

and

$$E\left\{\frac{1}{n}\sum_{j=1}^{n}\left(\frac{v_{n,j}^4}{s(\underline{x}_j)^4} - \frac{Ev_{n,j}^4}{s(\underline{x}_j)^4}\right)\right\}^2 = \frac{1}{n^2}\sum_{j=1}^{n}\frac{E(v_{n,j}^4 - Ev_{n,j}^4)^2}{s(\underline{x}_j)^8}$$

$$\leq \frac{1}{n.M^8}\int_{-a_n}^{a_n} u^8 dF(u) \leq \frac{1}{M^8}\,\frac{a_n^8}{n} \to 0 \text{ as } n\to\infty, \quad (3.4.19)$$

while

$$\text{plim}\,\frac{1}{n}\sum_{j=1}^{n}\left(\frac{Ev_{j.n}^4}{s(\underline{x}_j)^4} - E\left[\frac{v_{j,n}^4}{s(\underline{x}_j)^4}\right]\right) = \lim_{n\to\infty}\int_{-a_n}^{a_n} u^4 dF(u).\text{plim}\,\frac{1}{n}\sum_{j=1}^{n}\left\{\left(\frac{1}{s(\underline{x}_j)}\right)^4 - E\left(\frac{1}{s(\underline{x}_j)}\right)^4\right\} = 0. \quad (3.4.20)$$

Using Chebishev's inequality, we thus conclude

$$\text{plim}\left\{\frac{1}{n}\sum_{j=1}^{n}\left(\frac{u_j}{s(\underline{x}_j)}\right)^4 - \frac{1}{n}\sum_{j=1}^{n}E\left(\frac{u_j}{s(\underline{x}_j)}\right)^4\right\} = 0\,. \quad (3.4.21)$$

Consequently, since

$$\frac{1}{n}\sum_{j=1}^{n}E\left(\frac{u_j}{s(\underline{x}_j)}\right)^4 = \frac{1}{n}\sum_{j=1}^{n}\int u^4 f(u)du\int\left(\frac{1}{s(x)}\right)^4 dH_j(x) \to \int u^4 dF(u)\int\left(\frac{1}{s(x)}\right)^4 dH(x)\,, \quad (3.4.22)$$

we now see that (3.4.17) holds. Since under the conditions of theorem 3.1.1

$$\text{plim}\,\frac{1}{n}\sum_{j=1}^{n}\left(\frac{1}{s(\underline{x}_j)}\right)^4 = \int\left(\frac{1}{s(x)}\right)^4 dH(x)\,, \quad (3.4.23)$$

we have proved by now :

<u>Lemma 3.4.2.</u> Under the conditions of theorem 3.1.1 and the additional condition $\int u^4 dF(u) < \infty$ we have $\text{plim}\,\dfrac{\frac{1}{n}\sum_{j=1}^{n}\left(\frac{\hat{u}_j}{s(\underline{x}_j)}\right)^4}{\frac{1}{n}\sum_{j=1}^{n}\left(\frac{1}{s(\underline{x}_j)}\right)^4} = \int u^4 dF(u)$ for any continuous function $s(x)$ satisfying (3.4.14), where $\hat{u}_j = y_j - g(\underline{x}_j, \hat{\underline{\theta}}_n)$ and $\hat{\underline{\theta}}_n$ is a consistent estimator of θ_0(not necessarily NLLSE)

Thus if we denote the "weighted residual kurtosis" (WRK) by

$$\text{WRK} = \frac{\frac{1}{n}\sum_{j=1}^{n}\{\hat{u}_j/s(\underline{x}_j)\}^4}{\{\frac{1}{n}\sum_{j=1}^{n}\hat{u}_j^2\}^2\,\frac{1}{n}\sum_{j=1}^{n}\{\frac{1}{s(\underline{x}_j)}\}^4} - 3 \quad (3.4.24)$$

then under the conditions of the lemma, WRK converges in pr. to the kurtosis of the error distribution:

$$\text{WRK} \to \frac{\int u^4 dF(u)}{(\int u^2 dF(u))^2} - 3 \quad \text{in pr.} \tag{3.4.25}$$

For the case that $\int u^4 dF(u)=\infty$ we would expect that plim WRK=∞, and indeed this is true. For proving this we have to distinguish two cases, namely

$$\int |u|^3 dF(u)<\infty \text{ and } \int |u|^4 dF(u)=+\infty \tag{3.4.26}$$

and

$$\int |u|^3 dF(u)=+\infty \text{ (and consequently } \int |u|^4 dF(u)=+\infty) , \tag{3.4.27}$$

respectively. For the case (3.4.26) it follows from (3.4.15) that (3.4.16) still holds, while

$$\frac{1}{n}\sum_{j=1}^n \left(\frac{\underline{u}_j}{s(\underline{x}_j)}\right)^4 \geq \frac{1}{n}\sum_{j=1}^n \left(\frac{\underline{w}_j(A)}{s(\underline{x}_j)}\right)^4 \to \{\int_{-A}^{A} u^4 dF(u) + A^4\int_A^\infty dF(u) + A^4\int_{-\infty}^{-A} dF(u)\}\int \left(\frac{1}{s(x)}\right)^4 dH(x) \quad \text{in pr.} \tag{3.4.28}$$

for any A>0, where

$$\underline{w}_j(A)=\begin{cases} \underline{u}_j & \text{if } |\underline{u}_j|\leq A \\ -A & \text{if } \underline{u}_j<-A \\ A & \text{if } \underline{u}_j>A \end{cases} . \tag{3.4.29}$$

Hence

$$\text{plim } \frac{1}{n}\sum_{j=1}^n \left(\frac{\underline{u}_j}{s(\underline{x}_j)}\right)^4 =+\infty , \tag{3.4.30}$$

and consequently by (3.4.15)

$$\text{plim } \frac{1}{n}\sum_{j=1}^n \left(\frac{\hat{\underline{u}}_j}{s(\underline{x}_j)}\right)^4 =+\infty . \tag{3.4.31}$$

For the case (3.4.27) we note that the last term in the right side of inequality (3.4.15) may be replaced by $4|\hat{\underline{\theta}}_n-\theta_0|K \frac{1}{n}\sum_{j=1}^n \left|\frac{\underline{u}_j}{s(\underline{x}_j)}\right|^3$, hence it follows that

$$\text{plim } \frac{\frac{1}{n}\sum_{j=1}^n \left(\frac{\hat{\underline{u}}_j}{s(\underline{x}_j)}\right)^4 - \frac{1}{n}\sum_{j=1}^n \left(\frac{\underline{u}_j}{s(\underline{x}_j)}\right)^4}{\frac{1}{n}\sum_{j=1}^n \left|\frac{\underline{u}_j}{s(\underline{x}_j)}\right|^3} =0 . \tag{3.4.32}$$

Moreover, by Liapounov's inequality we have

$$\frac{\frac{1}{n}\sum_{j=1}^n \left(\frac{\underline{u}_j}{s(\underline{x}_j)}\right)^4}{\frac{1}{n}\sum_{j=1}^n \left|\frac{\underline{u}_j}{s(\underline{x}_j)}\right|^3} \geq \{\frac{1}{n}\sum_{j=1}^n \left|\frac{\underline{u}_j}{s(\underline{x}_j)}\right|^3\}^{\frac{1}{3}} ,$$

so that similar to (3.4.30),

$$\operatorname{plim} \frac{\frac{1}{n}\sum_{j=1}^{n}\left(\frac{u_j}{s(\underline{x}_j)}\right)^4}{\frac{1}{n}\sum_{j=1}^{n}\left|\frac{u_j}{s(\underline{x}_j)}\right|^3} = +\infty \tag{3.4.33}$$

and consequently by (3.4.32),

$$\operatorname{plim} \frac{\frac{1}{n}\sum_{j=1}^{n}\left(\frac{\hat{u}_j}{s(\underline{x}_j)}\right)^4}{\frac{1}{n}\sum_{j=1}^{n}\left|\frac{u_j}{s(\underline{x}_j)}\right|^3} = +\infty \, . \tag{3.4.34}$$

Since $\operatorname{plim} \frac{1}{n}\sum_{j=1}^{n}\left|\frac{u_j}{s(x_j)}\right|^3 = +\infty$, (3.4.35) implies that (3.4.31) also holds for the case (3.4.27). So we have proved:

<u>Lemma 3.4.3.</u> Lemma 3.4.2 carries over if $\int u^4 dF(u) = \infty$

We now conclude:

<u>Theorem 3.4.3</u>. Let the conditions of theorem 3.1.1 be satisfied. Define the weighted residual kurtosis WRK by (3.4.24), where s(x) is any continuous function satisfying (3.4.14). Then plim WRK exists. If this probability limit is positive then the robust M-estimation method of section 3.2 is asymptotically more efficient than the least squares estimation method*).

<u>Remark</u>: We recall that if the error distribution is normal $N(0,\sigma^2)$ then plim WRK=0 and if $\int u^4 dF(u) = \infty$ then plim WRK=∞.

So the statistic WRK may serve as a decision criterion for deciding whether or not least squares estimates can be improved by robust M-estimation, provided of course that the error distribution is symmetric unimodal or satisfies assumption 3.4.1 and condition (3.4.12).

---

*) Because then the function $h(\frac{1}{\lambda})$ is decreasing in a neighborhood of 0. However, it might be possible that $h(\frac{1}{\lambda})$ is increasing in a neighborhood of 0 but decreasing below the level $\sigma^2 = \int u^2 dF(u)$ for larger λ. Thus the WRK-criterion is sufficient, but not automatically necessary.

4 NONLINEAR STRUCTURAL EQUATIONS

The most fundamental assumption of the nonlinear regression model is that for each observation the regressors are independent of the error term. If this assumption is not satisfied, the consistency of least squares estimators and robust M-estimators breaks down. Only in the linear case the assumption involved may be weakened by assuming that the error and the regressors are uncorrelated. But in many situations these assumptions cannot be made, for example when the regressors are subject to errors of measurement (errors in variables) or when the equation involved is a member of a simultaneous equation system where one or more of the explanatory variables are dependent variables of other equations of the system.

For linear structural equations, econometric theory is well developed but only a few authors have devoted attention to nonlinear structural equations. Assuming normality of the errors, Eisenpress and Greenstadt (1966) derived the full information maximum likelihood (FIML) estimator of the parameters of a set of nonlinear explicit structural equations; their contribution focusses on computation of the estimates involved.
Berndt, Hall, Hall and Hausman (1974) and Jorgenson and Laffont (1974) also consider FIML, but they compare it with an extension of Amemiya(1974, 1975)'s nonlinear two-stage least squares estimator. The former authors adopt Amemiya's approach without criticism. Amemiya claims that his approach is more general than that of Kelejian (1971) who was dealing with econometric systems linear in the parameters but nonlinear in the endogenous variables. But for showing consistency of his estimator, Amemiya needs assumptions of the type $\frac{1}{n}\sum_{j=1}^{n} w_j(\partial/\partial\theta)g(\underline{x}_j,\theta) \to G$ in pr. uniformly in θ, where G is a <u>constant</u> matrix, where the w_j's are vectors of instrumental variables and the model involved is (3.1.1), (but now without assuming that $\underline{u}_j$ and $\underline{x}_j$ are independent). But such assumptions are only realistic if $g(x,\theta)$ is linear in θ, otherwise counter examples are easily found. Gallant (1977) has extended the work of Amemiya and Jorgenson and Laffont to a system of nonlinear implicit equations, and he has probably drawn the same conclusion as we did, because his approach is based on a variant of Jennrich's tailnorm concept.

Since this study deals exclusively with single equation methods, only Amemiya's nonlinear two-stage least squares estimator will be discussed in section 4.1. However, in view of the above mentioned problem with his assumptions, we shall restate the theory. Moreover, following Gallant (1977) we shall mainly be concerned with implicit structural equations, because of generality as well as simplicity of notation.

In the sections 4.2 and 4.3 we shall introduce a new estimation method for simultaneous equations, based on the assumption that the errors are symmetrically distributed. The advantages of this method are that it requires less instrumental variables than the nonlinear two stage least squares method and that no assumptions about finiteness of moments of the error distribution are necessary.

Finally, section 4.3 is devoted to problems concerning the calculation of the estimator, the objective function and the asymptotic variance matrix of this method. Moreover, it will be shown that the optimal value of the objective function can be used as a test criterion for the symmetry assumption.

4.1 Nonlinear two-stage least squares

4.1.1 Introduction

Consider model (3.1.1) but now drop the assumption that $\underline{u}_j$ and $\underline{x}_j$ are mutually independent. Then model (3.1.1) may be considered as a nonlinear explicit structural equation, where some components of $\underline{x}_j$ are dependent variables of other structural equations of the system. The implicit form of this structural equation is:

$$r(\underline{z}_j,\theta_0)=\underline{u}_j \ , \quad j=1,2,\ldots,n,\ldots \ , \tag{4.1.1}$$

where:

$$\underline{z}_j'=(\underline{y}_j,\underline{x}_j') \ , \ r(\underline{z}_j,\theta)=\underline{y}_j-g(\underline{x}_j,\theta) . \tag{4.1.2}$$

For the sake of generality we shall work further with the implicit form (4.1.1), considering (4.1.2) as an example of such models. Thus, similar to assumption 3.1.1, we now assume:

Assumption 4.1.1. The $\underline{z}_j$'s are independent p-component vectors of both endogenous and exogenous variables, the $\underline{u}_j$'s are independent and identically distributed errors terms, θ_0 is a parameter vector in a known compact subset Ⓗ of R^q and the function $r(z,\theta)$ and its partial derivatives $(\partial/\partial\theta_i)r(z,\theta)$ and $(\partial/\partial\theta_{i_1})(\partial/\partial\theta_{i_2})r(z,\theta),(i,i_1,i_2,=1,2,\ldots,q)$, are continuous on $R^p \times$ Ⓗ .

Moreover, we shall need a sequence $(\underline{w}_j)$ of vectors of instrumental variables, for example consisting of exogeneous components of $\underline{z}_j$ and exogenous variables of other equations of the system, or functions of these exogenous variables. The usual assumption is that:

Assumption 4.1.2. For each j the m-component vectors $\underline{w}_j$ of instrumental variables and the $\underline{u}_j$'s are mutually independent. The sequence of pairs $(\underline{z}_j,\underline{w}_j)$ is independent. Moreover, $m \geq q$.

Furthermore, throughout this section we shall assume that the error distribution satisfies the assumptions 3.1.2 and 3.1.10.

Consider the objective function:

$$\underline{S}_n(\theta)=\left(\frac{1}{n}\sum_{j=1}^{n} r(\underline{z}_j,\theta)\underline{w}_j\right)'\underline{W}_n^{-1}\left(\frac{1}{n}\sum_{j=1}^{n} r(\underline{z}_j,\theta)\underline{w}_j\right) , \tag{4.1.3}$$

where:

$$\underline{W}_n=\frac{1}{n}\sum_{j=1}^{n} \underline{w}_j\underline{w}_j' \quad . \tag{4.1.4}$$

Then the "modified" nonlinear two-stage least squares (NL2SLS)-estimator $\underline{\bar{\theta}}_n$, say, is a random vector obtained from:

$$\underline{S}_n(\underline{\bar{\theta}}_n)=\inf_{\theta\in \textcircled{H}} \underline{S}_n(\theta) \ , \quad \underline{\bar{\theta}}_n \in \textcircled{H} \quad \text{a.s.} \ . \tag{4.1.5}$$

We recall that if $\underline{W}_n$ is a.s. nonsingular, then it follows from lemma 3.1.2 that at least one solution of (4.1.5) is a random vector.

4.1.2 Strong consistency

As before, the strong consistency of the nonlinear two stage least squares estimator will be proved by showing that:

$$\sup_{\theta\in \textcircled{H}} |\underline{S}_n(\theta)- S(\theta)| \to 0 \text{ a.s.},$$

where $S(\theta)$ is a continuous function with a unique infimum at $\theta=\theta_0$. For this we shall need the following variant of assumption 3.1.3.

Assumption 4.1.3. The distribution functions $H_j(z,w)$ of the pairs (z_j, w_j), respectively, satisfy:

$$\frac{1}{n}\sum_{j=1}^{n} H_j(z,w)\to H(z,w) \text{ properly} .$$

Note that from theorem 2.2.10 and the above assumption it follows that also the means of the marginal distributions $H_j^{(1)}(z)$ and $H_j^{(2)}(w)$, say, converge properly : $\frac{1}{n}\sum_{j=1}^{n} H_j^{(1)}(z)\to H^{(1)}(z)$ properly and $\frac{1}{n}\sum_{j=1}^{n} H_j^{(2)}(w)\to H^{(2)}(w)$ properly , because for any continuous bounded function $\phi(w)$ on R^m it follows that:

$$\frac{1}{n}\sum_{j=1}^{n}\int\phi(w)dH_j^{(2)}(w)=\frac{1}{n}\sum_{j=1}^{n}\int\phi(w)dH_j(z,w)\to\int\phi(w)dH(z,w)=\int\phi(w)dH^{(2)}(w)$$

and a similar result holds for the $H_j^{(1)}(z)$'s.

First let us consider an element of the matrix $\underline{W}_n$. Since the instrumental variables vectors are assumed to be independent, it follows from the strong law of large numbers (theorem 2.2.3A) and part b of the following assumption:

Assumption 4.1.4.

a) $\sup_n \frac{1}{n}\sum_{j=1}^{n}E|w_{i,j}|^{2+\delta}<\infty$

and

b) $E\underline{w}_{i,j}^4=O(j^\mu)$

for i=1,2,...,m, and some $\delta>0$, $\mu<1$, respectively ,

where the $\underline{w}_{i,j}$'s are components of the vector $\underline{w}_j$, that

$$\frac{1}{n}\sum_{j=1}^{n}\{\underline{w}_{i_1,j}\,\underline{w}_{i_2,j}-E\underline{w}_{i_1,j}\,\underline{w}_{i_2,j}\}\to 0 \text{ a.s.} \tag{4.1.6}$$

Moreover, from theorem 2.2.15 and the assumptions 4.1.3 and 4.1.4a we obtain:

$$\frac{1}{n}\sum_{j=1}^{n}E\underline{w}_{i_1,j}\,\underline{w}_{i_2,j}\to\int w_{i_1}w_{i_2}dH^{(2)}(w)=\int w_{i_1}w_{i_2}dH(z,w)\ , \tag{4.1.7}$$

where the w_i's are now components of the vector w.

Consequently we thus have:

$$\underline{W}_n\to W \text{ a.s.}\ , \tag{4.1.8}$$

where W is the matrix with elements given by the right hand side of (4.1.7). Since the elements of W^{-1} are continuous functions of those of W, provided that

Assumption 4.1.5. The matrix W with elements $\int w_{i_1}w_{i_2}dH(z,w)$ is non singular

(and hence positive definite), it follows from (4.1.8) and theorem 2.2.5 that:

$$\underline{W}_n^{-1}\to W^{-1} \text{ a.s.}\ . \tag{4.1.9}$$

Next we consider $\frac{1}{n}\sum_{j=1}^{n}r(\underline{z}_j,\theta)\underline{w}_{i,j}$. Since from the inequalities (2.3.16) and

(2.3.17), $\overline{r(z,\theta)w_i}^{\,(H)} \leq \frac{1}{2}(\overline{r(z,\theta)}^{\,(H)})^2 + \frac{1}{2}w_i^2$, it follows from assumption 4.1.4, the following additional assumption:

Assumption 4.1.6.

a) $\sup_n \frac{1}{n}\sum_{j=1}^{n} E\{\overline{r(\underline{z}_j,\theta)}^{\,(H)}\}^{2+\delta} < \infty$ for some $\delta>0$

and

b) $E\{\overline{r(\underline{z}_j,\theta)}^{\,(H)}\}^4 = O(j^{\mu})$ for some $\mu < 1$,

and theorem 2.3.3 that:

$$\sup_{\theta\in(H)} \left|\frac{1}{n}\sum_{j=1}^{n} r(\underline{z}_j,\theta)\underline{w}_{i,j} - \mu_i(\theta)\right| \to 0 \text{ a.s. }, \tag{4.1.10}$$

where:

$$\mu_i(\theta) = \int r(z,\theta)w_i\,dH(z,w). \tag{4.1.11}$$

From (4.1.9), (4.1.10) and lemma 3.3.1 we now conclude:

$$\sup_{\theta\in(H)} |\underline{S}_n(\theta) - S(\theta)| \to 0 \text{ a.s.}^{*)} , \tag{4.1.12}$$

where:

$$S(\theta) = \mu(\theta)'W^{-1}\mu(\theta) \tag{4.1.13}$$

with:

$$\mu(\theta)' = (\mu_1(\theta),\ldots,\mu_m(\theta)) \tag{4.1.14}$$

Since $r(\underline{z}_j,\theta_0)=\underline{u}_j$ and $\underline{u}_j$ and $\underline{w}_j$ are mutually independent (assumption 4.1.2), $E\frac{1}{n}\sum_{j=1}^{n} r(\underline{z}_j,\theta_0)\underline{w}_{i,j} = 0$ for all n and consequently $\mu(\theta_0)$ is a zero vector, hence $S(\theta_0)=0$. But is θ_0 unique? Let us return to the explicit equation (4.1.2) and suppose that $g(x,\theta)=x'\theta$. Then

$$E\frac{1}{n}\sum_{j=1}^{n} r(\underline{z}_j,\theta)\underline{w}_j = E\frac{1}{n}\sum_{j=1}^{n}\underline{u}_j\underline{w}_j + E\frac{1}{n}\sum_{j=1}^{n}\underline{w}_j\underline{x}_j'(\theta_0-\theta) = \left(\frac{1}{n}\sum_{j=1}^{n}E\,\underline{w}_j\,\underline{x}_j'\right)(\theta_0-\theta)$$

so that in this case $\mu(\theta)$ is a zero vector if and only if $\theta=\theta_0$, provided that $\lim \frac{1}{n}\sum_{j=1}^{n}E\underline{w}_j\,\underline{x}_j'$ is a matrix with rank q. However, in the general case under review this identification condition can no longer be expressed in terms of the variables themselves, but only by the following more complex assumption.

---

*) Since $|\underline{S}_n(\theta) - S(\theta)|$ is a continuous function of θ and the observations $(\underline{z}_1,w_1),\ldots,(\underline{z}_n,w_2)$, it follows from theorem 2.3.1 that $\sup_{\theta\in(H)}|\underline{S}_n(\theta) - S(\theta)|$ is a continuous function of the observations and hence a random variable.

Assumption 4.1.7. (Identification condition). If $\theta \neq \theta_0$ then $\lim \frac{1}{n}\sum_{j=1}^{n} Er(\underline{z}_j,\theta)\underline{w}_j$ is a nonzero vector ,

which is also adopted by Gallant (1977). Obviously under this condition we have $S(\theta)=0$ if and only if $\theta=\theta_0$. From (4.1.12) and lemma 3.1.3 we now obtain

Theorem 4.1.1. Under the assumption 3.1.2 and 4.1.1 through 4.1.7, $\underline{\overline{\theta}}_n \to \theta_0$ a.s. .

4.1.3 Asymptotic normality

Also now we shall derive the limiting distribution of $\sqrt{n}(\underline{\overline{\theta}}_n-\theta_0)$ by means of a Taylor expansion of $(\partial/\partial\theta)\underline{S}_n(\underline{\overline{\theta}}_n)$ around the point θ_0, and we can be rather brief about it. Thus we conclude from assumption 3.1.6 and lemma 3.1.5 that

$$\sqrt{n}(\partial/\partial\theta)\underline{S}_n(\underline{\overline{\theta}}_n)\to 0' \text{ a.s.} \tag{4.1.15}$$

and then we set forth conditions such that $(\partial/\partial\theta)(\partial/\partial\theta')\underline{S}_n(\theta)$ converges in pr. uniformly on Ⓗ to some limit matrix and $\sqrt{n}(\partial/\partial\theta')\underline{S}_n(\theta_0)$ converges in distribution to a multivariate normal distribution with zero mean.So let us consider:

$$(\partial/\partial\theta)(\partial/\partial\theta')\underline{S}_n(\theta)=2\frac{1}{n}\sum_{j=1}^{n}(\partial/\partial\theta')r(\underline{z}_j,\theta)\underline{w}_j'\underline{W}_n^{-1}\frac{1}{n}\sum_{j=1}^{n}\underline{w}_j(\partial/\partial\theta)r(\underline{z}_j,\theta)$$

$$+ 2\frac{1}{n}\sum_{j=1}^{n}r(\underline{z}_j,\theta)\underline{w}_j'\underline{W}_n^{-1}\frac{1}{n}\sum_{j=1}^{n}\underline{w}_j(\partial/\partial\theta)(\partial/\partial\theta')r(\underline{z}_j,\theta)$$

$$= 2\underline{B}_n^0(\theta)+ 2\underline{C}_n^0(\theta), \tag{4.1.16}$$

say. Similar to (4.1.12) we see that if

Assumption 4.1.8.

a) $\sup_n \frac{1}{n}\sum_{j=1}^{n} E\{\overline{(\partial/\partial\theta)r(\underline{z}_j,\theta)}^{Ⓗ}\}^{2+\delta} < \infty$

and

b) $E\{\overline{(\partial/\partial\theta_i)r(\underline{z}_j,\theta)}^{Ⓗ}\}^4=O(j^{\mu})$

for $i=1,2,\ldots,q$ and some $\delta>0$, $\mu<1$, respectively, then under the same conditions as for (4.1.12)

$$\underline{B}_n^0(\theta) \rightarrow B^0(\theta) \text{ a.s. uniformly on } \circledH , \qquad (4.1.17)$$

where:

$$B^0(\theta)=(b^0_{i_1,i_2}(\theta)) \text{ with } b^0_{i_1,i_2}(\theta)=(\partial/\partial\theta_{i_1})\mu(\theta)'W^{-1}(\partial/\partial\theta_{i_2})\mu(\theta) \qquad (4.1.18)$$

and $(\partial/\partial\theta_i)\mu(\theta)$ denotes the vector of derivatives with respect to θ_i of the components of $\mu(\theta)$. Moreover, if

<u>Assumption 4.1.9.</u> $\sup_n \frac{1}{n}\sum_{j=1}^n \{\overline{(\partial/\partial\theta_{i_1})(\partial/\partial\theta_{i_2})r(\underline{z}_j,\theta)}^{\circledH}\}^{2+\delta} < \infty$

for $i_1,i_2=1,2,\ldots,q$ and some $\delta>0$,

then by theorem 2.3.4

$$\underline{C}_n^0(\theta) \rightarrow C^0(\theta) \text{ in pr. uniformly on } \circledH , \qquad (4.1.19)$$

where

$$C^0(\theta)=(c^0_{i_1,i_2}(\theta)) \text{ with } c^0_{i_1,i_2}(\theta)=\mu(\theta)'W^{-1}(\partial/\partial\theta_{i_1})(\partial/\partial\theta_{i_2})\mu(\theta) . \qquad (4.1.20)$$

Since $C^0(\theta_0)=0$ because $\mu(\theta_0)'=0'$, it follows now from (4.1.16), (4.1.17) and (4.1.19) and theorem 2.3.2 that for any consistent estimator $\underline{\theta}_n^*$ (with $\underline{\theta}_n^* \in \circledH$ a.s.),

$$(\partial/\partial\theta)(\partial/\partial\theta')\underline{S}_n(\underline{\theta}_n^*) \rightarrow 2B^0(\theta_0) \text{ in pr. .} \qquad (4.1.21)$$

Finally we consider

$$\sqrt{n}(\partial/\partial\theta)\underline{S}_n(\theta_0) = 2\frac{1}{\sqrt{n}}\sum_{j=1}^n \underline{u}_j\underline{w}_j'\underline{W}_n^{-1}\frac{1}{n}\sum_{j=1}^n \underline{w}_j(\partial/\partial\theta)r(\underline{z}_j,\theta_0) . \qquad (4.1.22)$$

We shall establish the asymptotic normality of (4.1.22) in two steps. First we observe that under the previous assumptions, similar to (4.1.17) and (4.1.12)

$$\underline{W}_n^{-1}\frac{1}{n}\sum_{j=1}^n \underline{w}_j(\partial/\partial\theta_i)r(\underline{z}_j,\theta_0) \rightarrow W^{-1}(\partial/\partial\theta_i)\mu(\theta_0) \text{ a.s.} \qquad (4.1.23)$$

Second, from the assumptions 3.1.2, 4.1.2 and 4.1.4 we see that for any vector $\zeta \in R^m$, the sequence $(\underline{u}_j\underline{w}_j'\zeta)$ satisfies the conditions of Liapounov's central limit theorem (theorem 2.4.6), provided that

$\lim \text{var}(\frac{1}{\sqrt{n}}\sum_{j=1}^n \underline{u}_j\underline{w}_j'\zeta)=\sigma^2\zeta'\lim\frac{1}{n}\sum_{j=1}^n E\underline{w}_j\underline{w}_j'\zeta$ exists, which does indeed because

$\lim \frac{1}{n}\sum_{j=1}^n E\underline{w}_j\underline{w}_j'=W$. So we conclude from the theorems 2.4.6 and 2.4.2

$$\frac{1}{\sqrt{n}}\sum_{j=1}^{n}\underline{u}_j\underline{w}_j \rightarrow N_q(0,\sigma^2 W) \text{ in distr. .} \tag{4.1.24}$$

From (4.1.23), (4.1.24) and theorem 2.2.14 we now see that

$$\sqrt{n}(\partial/\partial\theta')\underline{S}_n(\theta_0) \rightarrow N_q(0,4\sigma^2 B^0(\theta_0)) \text{ in distr. .} \tag{4.1.25}$$

From (4.1.15), (4.1.21) and (4.1.25) and again theorem 2.2.14 we conclude that

$$\sqrt{n}(\underline{\overline{\theta}}_n-\theta_0) \rightarrow N_q(0,\sigma^2 B^0(\theta_0)^{-1}) \text{ in distr. ,} \tag{4.1.26}$$

provided that:

Assumption 4.1.10. The matrix $B^0(\theta_0)$ is nonsingular.

Moreover, putting

$$\underline{Q}_n^0(\theta)=\frac{1}{n}\sum_{j=1}^{n}r(\underline{z}_j,\theta)^2 , \tag{4.1.27}$$

it follows from theorem 2.3.3 and the assumptions of theorem 4.1.1 that:

$$\sup_{\theta\in\Theta}|\underline{Q}_n^0(\theta)-\int r(z,\theta)^2 dH(z,w)| \rightarrow 0 \quad \text{a.s.} \tag{4.1.28}$$

and hence:

$$\underline{Q}_n^0(\underline{\overline{\theta}}_n)\rightarrow\int r(z,\theta_0)^2 dH(z,w)=\int u^2 dF(u)=\sigma^2 \text{ a.s. .} \tag{4.1.29}$$

Since obviously by theorem 2.3.2 and (4.1.17):

$$\underline{B}_n^0(\underline{\overline{\theta}}_n)\rightarrow B^0(\theta_0) \text{ a.s. ,} \tag{4.1.30}$$

we now have

$$\underline{Q}_n^0(\underline{\overline{\theta}}_n)\underline{B}_n^0(\underline{\overline{\theta}}_n)^{-1} \rightarrow \sigma^2 B^0(\theta_0)^{-1} \text{ a.s. ,} \tag{4.1.31}$$

which completes the proof of the following theorem:

Theorem 4.1.2. Under the assumptions of theorem 4.1.1 and the additional assumptions 3.1.6 and 4.1.8 through 4.1.10 we have:

$$\sqrt{n}(\underline{\bar{\theta}}_n-\theta_0)\to N_q(0,\sigma^2 B^0(\theta_0)^{-1}) \text{ in distr.}$$

and:

$$\underline{Q}_n^0(\underline{\bar{\theta}}_n)\underline{B}_n^0(\underline{\bar{\theta}}_n)^{-1} \to \sigma^2 B^0(\theta_0)^{-1} \text{ a.s.},$$

where $B^0(\theta_0)$, $\underline{B}_n^0(\theta)$ and $\underline{Q}_n^0(\theta)$ are defined by (4.1.18), (4.1.16) and (4.1.27), respectively.

4.1.4 Weak consistency

The parts a) and b) of the assumptions 4.1.4, 4.1.6 and 4.1.8 enabled us to use theorem 2.3.3. But if only the a) parts of these assumptions are satisfied we can still use theorem 2.3.4 in order to obtain weak consistency of the estimators $\underline{\bar{\theta}}_n$ and $\underline{B}_n^0(\underline{\bar{\theta}}_n)$. Thus:

Theorem 4.1.3. If in the theorems 4.1.1 and 4.1.2 only the a) parts of the assumptions 4.1.4, 4.1.6 and 4.1.8 are satisfied, then "a.s." becomes "in pr.".

4.2 Minimum information estimators: introduction

4.2.1 Lack of instruments

In the first stage of building a simultaneous equation system, the attention of the analist is often focussed on a single equation without having precise ideas about the structure of the other equations. Especially when the system is large, it then may be unclear which variables are exogenous and hence which variables are suitable instruments. But also in the final stage of model building the boundary between endogenous and exogenous variables may not always be sharp. For example, dealing with a macro economic model the variables with respect to governments behaviour are often assumed to be exogenous. But undoubtly most governments do react to the actual economic situation: if for example the unemployment rate increases, government may try to stimulate the economy by lowering tax rates or by raising government expenditure, hence such variables are in fact endogenous. Since then the

the model involved is incomplete, which means that there are more endogenous variables than structural equations, estimation of the structural equations by an instrumental variables method may be hampered by lack of instruments. An analogous situation occurs when the explanatory variables of a regression model are subject to errors of measurement, because then the relation between the observed explanatory variables and the real exogenous variables involved is unobservable.

In the following we shall present a new estimation method for such cases. This method requires less instrumental variables than there are parameters to be estimated: for deriving asymptotic normality we shall only need one instrumental variable, and consistency can even be obtained without using any instrumental variable at all. We shall therefore call our approach Minimum Information Estimation. Moreover, since consistency will be proved without making assumptions about finiteness of moments of errors and regressors, this minimum information estimation method may be considered as robust.

4.2.2 Identification without using instrumental variables

First we show by an example how the parameters of an errors in variables model can be identified without using instrumental variables. Here identifiability of the parameters means that the distributions of the observations have certain characteristics which only hold for the true parameters [see Malinvaud (1970), chapter 2, section 10]. Thus we consider the simple linear regression model:

$$\underline{y}_j=\theta_1^0+\theta_2^0 x_j+\underline{u}_j\ ,\ \underline{u}_j\sim N(0,\sigma_u^2),\ j=1,2,\ldots,n,\ldots,$$

where the x_j's can only be observed with errors:

$$\underline{w}_j=x_j+\underline{v}_j,\ \underline{v}_j\sim N(0,\sigma_v^2).$$

Moreover, we assume that $\underline{u}_j$ and $\underline{v}_j$ are mutually independent, so that:

$$\begin{pmatrix}\underline{u}_j\\ \underline{v}_j\end{pmatrix}\sim N_2\left[\begin{pmatrix}0\\0\end{pmatrix},\begin{pmatrix}\sigma_u^2 & 0\\ 0 & \sigma_v^2\end{pmatrix}\right].$$

Putting

$$\underline{z}_j'=(\underline{y}_j,\underline{w}_j),\theta'=(\theta_1,\theta_2),\ r(\underline{z}_j,\theta)=\underline{y}_j-\theta_1-\theta_2\underline{w}_j\ ,$$

we then see that

$$r(\underline{z}_j,\theta)\sim N[(\theta_1^0-\theta_1)+(\theta_2^0-\theta_2)x_j,\sigma_u^2+\theta_2^2\sigma_v^2]$$

and consequently

$$E \sin(t.r(\underline{z}_j,\theta))=\sin(t((\theta_1^0-\theta_1)+(\theta_2^0-\theta_2)x_j))e^{-\frac{1}{2}t^2(\sigma_u^2+\theta_2^2\sigma_v^2)} .$$

Now let $G_n(x)$ be the empirical distribution of $\{x_1,...,x_n\}$ and suppose that

$G_n(x) \to G(x)$ properly.

Then from theorem 2.2.10

$$\frac{1}{n}\sum_{j=1}^n E \sin(t.r(\underline{z}_j,\theta)) \to e^{-\frac{1}{2}t^2(\sigma_u^2+\theta_2^2\sigma_v^2)}.\int \sin(t.((\theta_1^0-\theta_1)+(\theta_2^0-\theta_2)x))dG(x) =$$

$$=s(t,\theta),$$

say. Obviously $s(t,\theta_0)=0$ for all t and from theorem 2.4.3 it follows that θ_0 is the only point with this property, provided that $G(x-\alpha)$ is non symmetric for all α. Thus in this case the parameter vector θ_0 is identifiable, because only if $\theta=\theta_0$ then the distributions of the observations $\underline{z}_j$ are such that $\frac{1}{n}\sum_{j=1}^n P(r(\underline{z}_j,\theta)\leq u)$ converges properly to a symmetric distribution function $H(u)$, say.

Generally, adopting model(4.1.1)and assumption 4.1.1 and assuming in addition that

Assumption 4.2.1. The error distribution $F(u)$ is symmetric,

and

Assumption 4.2.2. The distribution functions $G_j(z)$, say, of the $\underline{z}_j$'s satisfy: $\frac{1}{n}\sum_{j=1}^n G_j(z)\to G(z)$ properly,

the parameter vector θ_0 is now said to be identifiable if

Assumption 4.2.3. The distribution function

$$H_\theta(u)=\int_{\{r(z,\theta)\leq u\}} dG(z)$$

is symmetric if and only if $\theta=\theta_0$.

Then

$$s(t,\theta)=\lim \frac{1}{n}\sum_{j=1}^n E \sin(t.r(\underline{z}_j,\theta))=\int\sin(t.r(z,\theta))dG(z)=\int\sin(t.u)dH_\theta(u) \quad (4.2.1)$$

is zero everywhere in t if and only if $\theta=\theta_0$, which identifies θ_0.

4.2.3 Consistent estimation without using instrumental variables

The above property of the function $s(t,\theta)$ implies that for any function $\phi(t)$ satisfying

$$\phi(t) > 0 \text{ on } (0,\infty), \int_0^\infty \phi(t)dt < \infty, \tag{4.2.2}$$

the function $S^*(\theta)$ defined by:

$$S^*(\theta)=\int_0^\infty s(t,\theta)^2\phi(t)dt \tag{4.2.3}$$

has a unique minimum (equal to zero) at $\theta=\theta_0$. This suggests that if we put:

$$\underline{S}_n^*(\theta)=\int_0^\infty \underline{s}_n(t,\theta)^2\phi(t)dt, \tag{4.2.4}$$

where:

$$\underline{s}_n(t,\theta)=\frac{1}{n}\sum_{j=1}^n \sin(t.r(\underline{z}_j,\theta)), \tag{4.2.5}$$

then the statistic $\underline{\theta}_n^0$ obtained from:

$$\underline{S}_n^*(\underline{\theta}_n^0)=\inf_{\theta\in Ⓗ} \underline{S}_n^*(\theta) \quad , \quad \underline{\theta}_n^0 \in Ⓗ \quad \text{a.s.} \tag{4.2.6}$$

may be a strongly consistent estimator of θ_0. For proving this, it suffices to show that $\underline{S}_n^*(\theta)\to S^*(\theta)$ a.s. pseudo-uniformly on Ⓗ, because then it follows from lemma 3.1.3 that $\underline{\theta}_n^0 \to \theta_0$ a.s..

However, since $|\underline{S}_n^*(\theta)- S^*(\theta)|$ is a continuous function of θ and the observations, $\sup_{\theta\in Ⓗ}|\underline{S}_n^*(\theta)- S^*(\theta)|$ is by theorem 2.3.1 a continuous function of the observations and hence a random variable. Thus $\underline{S}_n^*(\theta)\to S^*(\theta)$ a.s. pseudo-uniformly on Ⓗ implies in this case that

$$\sup_{\theta\in Ⓗ}|\underline{S}_n^*(\theta)- S^*(\theta)|\to 0 \quad \text{a.s..} \tag{4.2.7}$$

Although condition (4.2.2) fits the requirements for proving strong consistency we shall choose $\phi(t)$ such that

<u>Assumption 4.2.4.</u> $\phi(t) > 0$ everywhere on $(0,\infty)$ and $\int_0^\infty t^2\phi(t)dt < \infty$

because this will be needed for deriving the limiting distribution of $\underline{\theta}_n^0$.

For showing (4.2.7) we observe that from theorem 2.3.3 and the assumptions 4.1.1 and 4.2.2:

$$\sup_{(t,\theta)\in T\times Ⓗ} |\underline{s}_n(t,\theta)- s(t,\theta)| \to 0 \quad \text{a.s.} \tag{4.2.8}$$

for any compact interval T, and consequently, since both $\underline{s}_n(t,\theta)$ and $s(t,\theta)$ are uniformly bounded by 1,

$$\sup_{(t,\theta)\in T \times \textcircled{H}} |\underline{s}_n(t,\theta)^2 - s(t,\theta)^2| \to 0 \text{ a.s.}. \tag{4.2.9}$$

For any $\varepsilon > 0$ we can choose $T_\varepsilon = [-M_\varepsilon, M_\varepsilon]$ such that $2\int_{M_\varepsilon}^{\infty} \phi(t)dt < \varepsilon$ because ϕ is integrable. Hence

$$\sup_{\theta\in\textcircled{H}} |\underline{S}_n^*(\theta) - S^*(\theta)| \le \int_0^\infty \sup_{\theta\in\textcircled{H}} |\underline{s}_n(t,\theta)^2 - s(t,\theta)^2|\phi(t)dt \le$$

$$\le \sup_{(t,\theta)\in T_\varepsilon \times \textcircled{H}} |\underline{s}_n(t,\theta)^2 - s(t,\theta)^2| \int_0^\infty \phi(t)dt + \varepsilon . \tag{4.2.10}$$

Combining (4.2.9) and (4.2.10) we easily obtain (4.2.7). So we have proved by now:

> Theorem 4.2.1. Under the assumptions 4.1.1 and 4.2.1 through 4.2.4 it follows that the statistic $\underline{\theta}_n^0$ obtained from (4.2.6) is strongly consistent: $\underline{\theta}_n^0 \to \theta_0$ a.s.

Since no instrumental variables are involved, this theorem shows that consistent estimation of structural equations without using instrumental variables is possible.

4.2.4 Asymptotic normality

For establishing the asymptotic normality of $\sqrt{n}(\underline{\theta}_n^0 - \theta_0)$ we first show that

$$\sqrt{n}(\partial/\partial\theta')\underline{S}_n^*(\theta_0) \to N_q(0, 4\Sigma_1(\theta_0)) \text{ in distr.} \tag{4.2.11}$$

for some matrix $\Sigma_1(\theta_0)$, and then we set forth conditions such that

$$(\partial/\partial\theta)(\partial/\partial\theta')\underline{S}_n^*(\theta) \to (\partial/\partial\theta)(\partial/\partial\theta')S^*(\theta) \text{ in pr. uniformly on } \textcircled{H} . \tag{4.2.12}$$

For convenience, put

$$\underline{c}_{n,i}(t,\theta) = (\partial/\partial\theta_i)\underline{s}_n(t,\theta) = \frac{1}{n}\sum_{j=1}^n t.\cos(t.r(\underline{z}_j,\theta))(\partial/\partial\theta_i)r(\underline{z}_j,\theta) . \tag{4.2.13}$$

From assumption 4.2.4 it follows that we may differentiate under the integral of (4.2.3):

$$(\partial/\partial\theta_i)\underline{S}_n^*(\theta) = 2\int_0^\infty \underline{s}_n(t,\theta)\underline{c}_{n,i}(t,\theta)\phi(t)dt . \tag{4.2.14}$$

Moreover, under the assumptions 4.2.2 and 4.1.8 it follows from theorem 2.3.3 that for i=1,2,...,q

$$\sup_{(t,\theta)\in T\times \textcircled{H}} |\underline{c}_{n,i}(t,\theta)- c_i(t,\theta)| \to 0 \quad \text{a.s.} \tag{4.2.15}$$

for any compact interval T, where

$$c_i(t,\theta)=t\int\cos(t.r(z,\theta))(\partial/\partial\theta_i)r(z,\theta)dG(z)\ . \tag{4.2.16}$$

We show now that

$$\text{plim}\ \{\sqrt{n}(\partial/\partial\theta_i)\underline{S}_n^*(\theta_0)- 2\int_0^\infty\sqrt{n}\ \underline{s}_n(t,\theta_0)c_i(t,\theta_0)\phi(t)dt\}=0\ , \tag{4.2.17}$$

because then $\sqrt{n}(\partial/\partial\theta)\underline{S}_n^*(\theta_0)$ converges in distribution to the normal if the random vector with components $2\int_0^\infty \sqrt{n}\ \underline{s}_n(t,\theta_0)c_i(t,\theta_0)\phi(t)dt\,(i=1,2,\ldots,q)$ does.

Observe that from Schwartz inequality, treating $\phi(t)$ like a density, that

$$|\int\sqrt{n}\ \underline{s}_n(t,\theta)\{\underline{c}_{n,i}(t,\theta)- c_i(t,\theta)\}\phi(t)dt| \le$$
$$\le [\int_0^\infty n\ \underline{s}_n(t,\theta)^2\phi(t)dt.\ \int_0^\infty\{\underline{c}_{n,i}(t,\theta)- c_i(t,\theta)\}^2\phi(t)dt]^{\frac{1}{2}} \quad \text{a.s.}, \tag{4.2.18}$$

while from theorem 2.2.5 we see that (4.2.15) implies

$$\sup_{(t,\theta)\in T\times \textcircled{H}} \{\underline{c}_{n,i}(t,\theta)- c_i(t,\theta)\}^2 \to 0 \quad \text{a.s.} \tag{4.2.19}$$

and hence, similar to (4.2.10),

$$\int_0^\infty\{\underline{c}_{n,i}(t,\theta_0)- c_i(t,\theta_0)\}^2\ \phi(t)dt \to 0 \quad \text{a.s.}\ . \tag{4.2.20}$$

Furthermore it follows from the independence and the symmetry of the errors $\underline{u}_j$ that $E\ n\ \underline{s}_n(t,\theta_0)^2= \frac{1}{n}\sum_{j=1}^n E\ \sin(t\underline{u}_j)^2 \le 1$ and hence

$$E\int_0^\infty n\ \underline{s}_n(t,\theta_0)^2\phi(t)dt=\int_0^\infty n\ E\ \underline{s}_n(t,\theta)^2\phi(t)dt \le \int_0^\infty\phi(t)dt\ . \tag{4.2.21}$$

Consequently we conclude from Chebishev's inequality that the sequence of random variables

$$\underline{v}_n=\int_0^\infty n\ \underline{s}_n(t,\theta_0)^2\phi(t)dt$$

is stochastically bounded, which means that for any $\varepsilon > 0$ we can find a number $K_\varepsilon < \infty$ such that

$$P(|\underline{v}_n| \leq K_\varepsilon) > 1-\varepsilon \quad \text{for all } n .$$

But

Lemma 4.2.1. If $\underline{v}_n$ is stochastically bounded and $\underline{w}_n \to 0$ in pr., then $\underline{v}_n \underline{w}_n \to 0$ in pr.,

because for any $\varepsilon > 0$ and any $\delta > 0$,

$$P(|\underline{v}_n\underline{w}_n|>\varepsilon)=P(|\underline{v}_n\underline{w}_n|>\varepsilon \text{ and } |\underline{v}_n| \leq K_\delta)+P(|\underline{v}_n\underline{w}_n|>\varepsilon \text{ and } |\underline{v}_n| > K_\delta)$$

$$\leq P(|\underline{w}_n| > \frac{\varepsilon}{K_\delta})+ P(|\underline{v}_n|>K_\delta) \leq 2\delta$$

for n sufficiently large. Thus (4.2.20) and (4.2.21) imply that the right side of (4.2.18) converges to zero in probability, and hence that (4.2.17) holds. Furthermore, (4.2.17) and theorem 2.4.2 imply that for showing (4.2.11) it suffices to show that

$$\sqrt{n}\int_0^\infty \underline{s}_n(t,\theta_0)\zeta' c(t,\theta_0)\phi(t)dt = \frac{1}{\sqrt{n}}\sum_{j=1}^n \int_0^\infty \sin(t\underline{u}_j)\zeta' c(t,\theta_0)\phi(t)dt$$

$$= \frac{1}{\sqrt{n}}\sum_{j=1}^n \underline{x}_j, \tag{4.2.22}$$

say, converges in distribution to $N(0,\xi'\Sigma_1(\theta_0)\xi)$, where

$$c(t,\theta)'=(c_1(t,\theta),\ldots,c_q(t,\theta)) \tag{4.2.23}$$

and ζ is any vector in R^k. Since the $\underline{x}_j$'s are independent and have zero expectation (because $E\sin(t\underline{u}_j)=0$), it follows that:

$$\text{var}(\frac{1}{\sqrt{n}}\sum_{j=1}^n \underline{x}_j)= \frac{1}{n}\sum_{j=1}^n E\underline{x}_j^2=\zeta'\,\Sigma_1(\theta_0)\zeta \quad , \tag{4.2.24}$$

where:

$$\Sigma_1(\theta)=(\int_0^\infty\int_0^\infty \psi(t_1,t_2)c_{i_1}(t_1,\theta)c_{i_2}(t_2,\theta)\phi(t_1)\phi(t_2)dt_1dt_2), \tag{4.2.25}$$

with:

$$\psi(t_1,t_2)=\int \sin(t_1u)\sin(t_2u)dF(u). \tag{4.2.26}$$

It is easily verified now that if $\zeta'\Sigma_1(\theta_0)\zeta > 0$ then the conditions of theorem 2.4.6 are satisfied, so that

$$\frac{1}{\sqrt{n}}\sum_{j=1}^n \underline{x}_j \rightarrow N(0,\zeta'\Sigma_1(\theta_0)\zeta) \text{ in distr..}$$

Moreover, if for some nonzero vector ζ, $\zeta'\Sigma_1(\theta_0)\zeta = 0$, which is possible if $\Sigma_1(\theta_0)$ is positive semidefinite, then (4.2.24) implies that $\frac{1}{\sqrt{n}}\sum_{j=1}^n \underline{x}_j \rightarrow 0$ in pr. which however may also be intepreted as

$$\frac{1}{\sqrt{n}}\sum_{j=1}^n \underline{x}_j \rightarrow N(0,0)=N(0,\zeta'\Sigma_1(\theta_0)\zeta) \text{ in distr..}$$

So we conclude now that (4.2.11) holds (where the normal distribution involved may be singular).

Next we turn to the proof of (4.2.12). Consider

$$(\partial/\partial\theta_{i_1})(\partial/\partial\theta_{i_2})\underline{S}_n^*(\theta)=2\int_0^\infty\{(\partial/\partial\theta_{i_1})\underline{s}_n(t,\theta)\}\{(\partial/\partial\theta_{i_2})\underline{s}_n(t,\theta)\}\phi(t)dt +$$
$$+2\int_0^\infty \underline{s}_n(t,\theta)(\partial/\partial\theta_{i_1})(\partial/\partial\theta_{i_2})\underline{s}_n(t,\theta)\phi(t)dt$$
$$=2\int_0^\infty \underline{c}_{n,i_1}(t,\theta)\underline{c}_{n,i_2}(t,\theta)\phi(t)dt + 2\int_0^\infty \underline{s}_n(t,\theta)\underline{d}_{n,i_1,i_2}(t,\theta)\phi(t)dt, \quad (4.2.27)$$

where:

$$\underline{d}_{n,i_1,i_2}(t,\theta)=(\partial/\partial\theta_{i_1})(\partial/\partial\theta_{i_2})\underline{s}_n(t,\theta) =$$
$$=-\frac{1}{n}\sum_{j=1}^n t^2\sin(t.r(\underline{z}_j,\theta))\{(\partial/\partial\theta_{i_1})r(\underline{z}_j,\theta)\}\{(\partial/\partial\theta_{i_2})r(\underline{z}_j,\theta)\}$$
$$+\frac{1}{n}\sum_{j=1}^n t\cos(t.r(\underline{z}_j,\theta))(\partial/\partial\theta_{i_1})(\partial/\partial\theta_{i_2})r(\underline{z}_j,\theta) \, . \quad (4.2.28)$$

Under assumption 4.1.8a and

<u>Assumption 4.2.5.</u> $\sup_n \frac{1}{n}\sum_{j=1}^n E\{\overline{(\partial/\partial\theta_{i_1})(\partial/\partial\theta_{i_2})r(\underline{z}_j,\theta)}^{(H)}\}^{1+\delta} < \infty$

for $i_1,i_2=1,2,\ldots,q$ and some $\delta>0$,

it follows from theorem 2.3.4 that

$$\text{plim} \sup_{(t,\theta)\in T\times (H)} |\underline{d}_{n,i_1,i_2}(t,\theta)-d_{i_1,i_2}(t,\theta)|=0 \quad (4.2.29)$$

for any compact interval T, where

$$d_{i_1,i_2}(t,\theta)=-t^2\int \sin(t.r(z,\theta))\{(\partial/\partial\theta_{i_1})r(z,\theta)\}\{(\partial/\partial\theta_{i_2})r(z,\theta)\}dG(z)$$
$$+\ t\int \cos(t.r(z,\theta))(\partial/\partial\theta_{i_1})(\partial/\partial\theta_{i_2})r(z,\theta)dG(z) \qquad (4.2.30)$$

Consequently, similar to (4.2.10) it follows now from (4.2.9), (4.2.29) and assumption 4.2.4 that:

$$\text{plim}\sup_{\theta\in Ⓗ}\left|\int \underline{s}_n(t,\theta)\underline{d}_{n,i_1,i_2}(t,\theta)\phi(t)dt-\int_0^\infty s(t,\theta)d_{i_1,i_2}(t,\theta)\phi(t)dt\right|=0 \qquad (4.2.31)$$

Furthermore, from (4.2.15) and assumption 4.2.4 we have

$$\text{plim}\sup_{\theta\in Ⓗ}\left|\int_0^\infty \underline{c}_{n,i_1}(t,\theta)\underline{c}_{n,i_2}(t,\theta)\phi(t)dt-\int_0^\infty c_{i_1}(t,\theta)c_{i_2}(t,\theta)\phi(t)dt\right|=0 \qquad (4.2.32)$$

Thus (4.2.12) holds, and consequently for any statistic $\underline{\theta}_n^*$ satisfying $\underline{\theta}_n^* \in Ⓗ$ a.s. and plim $\underline{\theta}_n^*=\theta_0$ we have by theorem 2.3.2

$$\text{plim}(\partial/\partial\theta)(\partial/\partial\theta')\underline{S}_n^*(\underline{\theta}_n^*)=(\partial/\partial\theta)(\partial/\partial\theta')S^*(\theta_0)=\Sigma_2(\theta_0), \qquad (4.2.33)$$

say, where

$$\Sigma_2(\theta)=(\int_0^\infty c_{i_1}(t,\theta)c_{i_2}(t,\theta)\phi(t)dt)\ . \qquad (4.2.34)$$

(Notice that we have used the fact that $\int_0^\infty s(t,\theta_0)d_{i_1,i_2}(t,\theta_0)\phi(t)dt=0$).

Under assumption 3.1.6 this statistic $\underline{\theta}_n^*$ can be chosen such that:

$$\sqrt{n}(\partial/\partial\theta)\underline{S}_n^*(\underline{\theta}_n^0)=\sqrt{n}(\partial/\partial\theta)\underline{S}_n^*(\theta_0)+\sqrt{n}(\underline{\theta}_n^0-\theta_0)'(\partial/\partial\theta)(\partial/\partial\theta')\underline{S}_n^*(\underline{\theta}_n^*)$$

and

$$|\underline{\theta}_n^*-\theta_0|\leq|\underline{\theta}_n^0-\theta_0|\ .$$

Since from assumption 3.1.6 and lemma 3.1.5, $\sqrt{n}(\partial/\partial\theta)\underline{S}_n^*(\underline{\theta}_n^0)\to 0'$ a.s., we now conclude

<u>Theorem 4.2.2.</u> Under the assumptions of theorem 4.2.1 and the additional assumptions 3.1.6, 4.1.8 and 4.2.5,

$$\sqrt{n}(\underline{\theta}_n^0-\theta_0)\to N_q[0,\Sigma_2^{-1}(\theta_0)\Sigma_1(\theta_0)\Sigma_2^{-1}(\theta_0)] \text{ in distr.}$$

provided that $\Sigma_2(\theta_0)$ is nonsingular.

4.2.5 A problem concerning the non-singularity assumption

However, a serious problem arises now: the matrix $\Sigma_2(\theta_0)$ is singular for linear homoscedastic models. For example, considering the linear errors in variables model discussed in section 4.2.2, we see that

$$c_1(t,\theta)=(\partial/\partial\theta_1)s(t,\theta)=-te^{-\frac{1}{2}t^2(\sigma_u^2+\theta_2^2\sigma_v^2)}\int\cos(t.((\theta_1^0-\theta_1)+(\theta_2^0-\theta_2)x))dG(x)$$

$$c_2(t,\theta)=(\partial/\partial\theta_2)s(t,\theta)=-t^2\sigma_v^2\,\theta_2 s(t,\theta)-$$

$$-\,te^{-\frac{1}{2}t^2(\sigma_u^2+\theta_2^2\sigma_v^2)}\int x\cos(t((\theta_1^0-\theta_1)+(\theta_2^0-\theta_2)x))dG(x)\ ,$$

and hence

$$\Sigma_2(\theta_0)=(\int_0^\infty c_{i_1}(t,\theta_0)c_{i_2}(t,\theta_0)\phi(t)dt)=\int_0^\infty t^2e^{-t^2(\sigma_u^2+(\theta_2^0)^2\sigma_v^2)}\phi(t)dt\begin{pmatrix}1 & ,\int xdG(x)\\ \int xdG(x), & (\int xdG(x))^2\end{pmatrix},$$

which is obviously a singular matrix.

Consider again the linear errors in variables model involved, but assume now that

$$\underline{u}_j\sim N(0,\sigma^2.x_j^2).$$

Then

$$r_j(\underline{z}_j,\theta)\sim N((\theta_1^0-\theta_1)+(\theta_2^0-\theta_2)x_j,\sigma^2x_j^2+(\theta_2)^2\sigma_v^2),$$

and thus after some calculations we obtain

$$c_1(t,\theta_0)=-t\int e^{-\frac{1}{2}t^2(\sigma^2.x^2+(\theta_2^0)^2\sigma_v^2)}dG(x),$$

$$c_2(t,\theta_0)=-t\int x.e^{-\frac{1}{2}t^2(\sigma^2.x^2+(\theta_2^0)^2\sigma_v^2)}dG(x).$$

If for example

$$G(x)=\frac{\sqrt{2}}{\sqrt{\pi}}\int_0^x e^{-\frac{1}{2}u^2}du\ , \qquad (4.2.35)$$

then

$$c_1(t,\theta_0)=-te^{-\frac{1}{2}t^2(\theta_2^0)^2\sigma_v^2}.\frac{\sqrt{2}}{\sqrt{\pi}}\int_0^\infty e^{-\frac{1}{2}(t^2\sigma^2+1)x^2}dx=$$

$$= \frac{-t}{\sqrt{\sigma^2 t^2+1}} e^{-\frac{1}{2}t^2(\theta_2^0)^2\sigma_v^2},$$

$$c_2(t,\theta_0)=te^{-\frac{1}{2}t^2(\theta_2^0)^2\sigma_v^2} \cdot \frac{\sqrt{2}}{\sqrt{\pi}}\int_0^\infty xe^{-\frac{1}{2}(t^2\sigma^2+1)x^2}dx$$

$$= \frac{\sqrt{2}}{\sqrt{\pi}} \cdot \frac{-t}{\sigma^2 t^2+1} e^{-\frac{1}{2}t^2(\theta_2^0)^2\sigma_v^2},$$

so that

$$\Sigma_2(\theta_0)=\begin{pmatrix} \int_0^\infty \frac{t^2}{\sigma^2t^2+1} e^{-t^2(\theta_2^0)\sigma_v^2}\phi(t)dt, & \frac{\sqrt{2}}{\sqrt{\pi}}\int_0^\infty \frac{t^2}{(\sigma^2t^2+1)^{3/2}} e^{-t^2(\theta_2^0)^2\sigma_v^2}\phi(t)dt \\ \frac{\sqrt{2}}{\sqrt{\pi}}\int_0^\infty \frac{t^2}{(\sigma^2t^2+1)^{3/2}} e^{-t^2(\theta_2^0)^2\sigma_v^2}\phi(t)dt, & \frac{2}{\pi}\int_0^\infty \frac{t^2}{(\sigma^2t^2+1)^2} e^{-t^2(\theta_2^0)^2\sigma_v^2}\phi(t)dt \end{pmatrix}$$

Putting $\psi(t)=\phi(t)t^2e^{-t^2(\theta_2^0)^2\sigma_v^2}/(k(\sigma^2t^2+1)^3)$, where k is chosen such that $\int_0^\infty \psi(t)dt=1$,

we now see that det $\Sigma_2(\theta_0)=0$ if

$$\int_0^\infty(\sigma^2t^2+1)^2\psi(t)dt\int_0^\infty(\sigma^2t^2+1)\psi(t)dt=\{\int_0^\infty(\sigma^2t^2+1)^{3/2}\ \psi(t)dt\}^2.$$

But from Hölder's inequality we always have

$$\{\int_0^\infty (\sigma^2t^2+1)^{3/2}\psi(t)dt\}^2 \leq \int_0^\infty(\sigma^2t^2+1)^2\psi(t)dt\int_0^\infty(\sigma^2t^2+1)\psi(t)dt ,$$

where only in special cases the sign "$\leq$" can be replaced with "=" [compare the proof of Hölder's inequality in section 2.1.4 and the fact that -log x is strictly convex on $(0,\infty)$]. Thus the assumption that $\Sigma_2(\theta_0)$ is nonsingular may be realistic in the heteroscedastic case. This example suggests the way out of the problem: Let $(\underline{w}_j)$ be a sequence of random variables such that $\underline{u}_j$ and $\underline{w}_j$ are independent but $\underline{z}_j$ and $\underline{w}_j$ are dependent. Then instead of the original implicit equation $r(\underline{z}_j,\theta_0)=\underline{u}_j$ we use the artificially disturbed model $\underline{w}_j r(\underline{z}_j,\theta_0)=\underline{w}_j\underline{u}_j$ so that the "new" disturbances $\underline{w}_j\underline{u}_j$ act like heteroscedastic random variables. In that case we may expect that the matrix $\Sigma_2(\theta_0)$ involved is nonsingular. The sequence $(\underline{w}_j)$ may be considered as a sequence of instrumental variables.

4.3 Minimum information estimators: instrumental variable and scaling parameter

4.3.1 An instrumental variable

The most appropriate choice of the instrumental variable $\underline{w}_j$ mentioned in the previous section 4.2.5 is to let it be some function of all the exogenous variables of the system because then $\underline{w}_j$ and $\underline{u}_j$ are surely independent since the exogenous variables are independent of $\underline{u}_j$, while obviously $\underline{w}_j$ is dependent of each component of $\underline{z}_j$. But it is not always necessary to take all the exogenous variables into account, and it may even be sufficient in particular cases to take only the exogenous components of $\underline{z}_j$. In any way we can add new components to the vector $\underline{z}_j$ such that $\underline{w}_j=w(\underline{z}_j)$ for some function $w(z)$. In view of theorem 2.3.3 it is convenient to choose $w(z)$ to be a continuous function. Thus:

> Assumption 4.3.1. $\underline{w}_j=w(\underline{z}_j)$ is used as instrumental variable, where w is a continuous function on R^p such that $w(\underline{z}_j)$ and $\underline{u}_j$ are mutually independent.

Then instead of the original implicit equation (4.1.1) we now shall work further with the artificially disturbed equation

$$r^*(\underline{z}_j,\theta_0)=w(\underline{z}_j)r(\underline{z}_j,\theta_0)=\underline{u}_j w(\underline{z}_j)=\underline{u}_j^*,\ j=1,2,\ldots \tag{4.3.1}$$

The only difference between the situation considered in the previous section 4.2 and the present one is that the new disturbances $\underline{u}_j^*$ are no longer homoscedastic. Note that the $\underline{u}_j^*$'s are still symmetrically distributed because $\underline{u}_j$ and $w(\underline{z}_j)$ are assumed to be independent.

The identification assumption 4.2.3 now becomes

> Assumption 4.3.2. The distribution function $H_\theta^*(u)=\int_{\{w(z)r(z,\theta)\leq u\}} dG(z)$ is symmetric if and only if $\theta=\theta_0$.

Under this assumption we obviously have that

$$s(t,\theta)=\lim\frac{1}{n}\sum_{j=1}^{n}E\sin(t.w(\underline{z}_j)r(\underline{z}_j,\theta))=\int\sin(t.w(z)r(z,\theta))dG(z)=\int\sin(t.u)dH_\theta^*(u) \tag{4.3.2}$$

is zero everywhere in t if and only if $\theta=\theta_0$. It is easily seen now that nearly the whole argument of section 4.2 carries over for the case (4.3.1), provided that assumption 4.3.1 is added and assumption 4.2.3 is replaced with assumption 4.3.2. Only the function $\psi(t_1,t_2)$ in (4.2.26) has to be redefined as follows:

$$\psi(t_1,t_2)=\lim_{n\to\infty}\frac{1}{n}\sum_{j=1}^{n}E\ \sin(t_1\underline{u}_j w(\underline{z}_j))\sin(t_2\underline{u}_j w(\underline{z}_j))$$

$$=\iint\sin(t_1.u.w(z))\sin(t_2.u.w(z))dF(u)dG(z)\ . \qquad (4.3.3)$$

4.3.2 An example

As an example of a model satisfying assumption 4.3.2 and the condition $\det\Sigma_2(\theta_0)\neq 0$ we consider the system

$$\begin{aligned}\underline{y}_{1,j}&=\theta_1^0\underline{y}_{2,j}+\theta_2^0 x_{1,j}+\underline{u}_{1,j}\ ,\\ \underline{y}_{2,j}&=\alpha\underline{y}_{1,j}+\beta\, x_{2,j}+\underline{u}_{2,j}\ ,\end{aligned} \qquad (4.3.4)$$

where the vectors $\begin{pmatrix}\underline{u}_{1,j}\\ \underline{u}_{2,j}\end{pmatrix}$ are random drawings from a bivariate normal distribution with zero mean vector and variance matrix equal to the unit matrix, and the $x_{1,j}$'s are non-random exogenous variables.

Put:

$$\underline{z}_j'=(\underline{y}_{1,j},\underline{y}_{2,j},x_{1,j}),\theta'=(\theta_1,\theta_2),r(\underline{z}_j,\theta)=\underline{y}_{1,j}-\theta_1\underline{y}_{2,j}-\theta_2 x_{1,j}\ , \qquad (4.3.5)$$

$$w(\underline{z}_j)=x_{1,j}\ .$$

Then:

$$\begin{pmatrix}\underline{y}_{1,j}\\ \underline{y}_{2,j}\end{pmatrix}=\frac{1}{1-\alpha\theta_1^0}\begin{pmatrix}1 & \theta_1^0\\ \alpha & 1\end{pmatrix}\begin{pmatrix}\theta_2^0 & 0\\ 0 & \beta\end{pmatrix}\begin{pmatrix}x_{1,j}\\ x_{2,j}\end{pmatrix}+\frac{1}{1-\alpha\theta_1^0}\begin{pmatrix}1 & \theta_1^0\\ \alpha & 1\end{pmatrix}\begin{pmatrix}\underline{u}_{1,j}\\ \underline{u}_{2,j}\end{pmatrix}$$

$$\sim N_2\left[\frac{1}{1-\alpha\theta_1^0}\begin{pmatrix}\theta_2^0 & \theta_1^0\beta\\ \alpha\theta_2^0 & \beta\end{pmatrix}\begin{pmatrix}x_{1,j}\\ x_{2,j}\end{pmatrix}\ ,\ \frac{1}{(1-\alpha\theta_1^0)^2}\begin{pmatrix}1+(\theta_1^0)^2, & \alpha+\theta_1^0\\ \alpha+\theta_1^0 & ,\ 1+\alpha^2\end{pmatrix}\right]\ , \qquad (4.3.6)$$

hence, after some calculations,

$$r(\underline{z}_j,\theta)\sim N\{[(\theta_2^0-\theta_2)+\frac{\alpha\theta_2^0}{1-\alpha\theta_1^0}(\theta_1^0-\theta_1)]x_{1,j}+\frac{(\theta_1^0-\theta_1)\beta}{1-\alpha\theta_1^0}x_{2,j},\ \sigma_\theta^2\}\ , \qquad (4.3.7)$$

where:

$$\sigma_\theta^2 = \frac{1}{(1-\alpha\theta_1^0)^2}(1,-\theta_1)\begin{pmatrix}1+(\theta_1^0)^2, & \alpha+\theta_1^0\\ \alpha+\theta_1^0 &, 1+\alpha^2\end{pmatrix}\begin{pmatrix}1\\ -\theta_1\end{pmatrix} . \tag{4.3.8}$$

From (4.3.7) and (4.3.5) we obtain

$$E\underline{s}_n(t,\theta)=\frac{1}{n}\sum_{j=1}^n E\ \sin(t.w(\underline{z}_j)r(\underline{z}_j,\theta))=$$

$$\frac{1}{n}\sum_{j=1}^n \sin(t[\{(\theta_2^0-\theta_2)+\frac{\alpha\theta_2^0}{1-\alpha\theta_1^0}(\theta_1^0-\theta_1)\}x_{1,j}^2+\frac{(\theta_1^0-\theta_1)\beta}{1-\alpha\theta_1^0}x_{1,j}\,x_{2,j}])e^{-\frac{1}{2}t^2\sigma_\theta^2x_{1,j}^2} \tag{4.3.9}$$

Now let $G_n(x_1,x_2)$ be the empirical distribution function of $\begin{pmatrix}x_{1,1}\\ x_{2,1}\end{pmatrix},\dots,\begin{pmatrix}x_{1,n}\\ x_{2,n}\end{pmatrix}$

and suppose that

$$G_n(x_1,x_2) \to G(x_1,x_2)=G_1(x_1)G_2(x_2) \quad \text{properly} , \tag{4.3.10}$$

where $G_1(x)$ equals the distribution (4.2.35):

$$G_1(x)=\frac{\sqrt{2}}{\sqrt{\pi}}\int_0^x e^{-\frac{1}{2}u^2}du , \tag{4.3.11}$$

and $G_2(x)$ is:

$$G_2(x)=\begin{cases}1 \text{ if } x\geq 1\\ 0 \text{ if } x<1 .\end{cases} \tag{4.3.12}$$

Then from theorem 2.2.10:

$$s(t,\theta)=\lim E\underline{s}_n(t,\theta)=$$

$$\iint \sin(t[\{(\theta_2^0-\theta_2)+\frac{\alpha\theta_2^0}{1-\alpha\theta_1^0}(\theta_1^0-\theta_1)\}x_1^2+\frac{(\theta_1^0-\theta_1)\beta}{1-\alpha\theta_1^0}x_1x_2])e^{-\frac{1}{2}t^2\sigma_\theta^2x_1^2}dG_1(x_1)dG_2(x_2)=$$

$$=\frac{\sqrt{2}}{\sqrt{\pi}}\int_0^\infty \sin(t[\{(\theta_2^0-\theta_2)+\frac{\alpha\theta_2^0}{1-\alpha\theta_1^0}(\theta_1^0-\theta_1)\}x^2+\frac{(\theta_1^0-\theta_1)\beta}{1-\alpha\theta_1^0}x])e^{-\frac{1}{2}(t^2\sigma_\theta^2+1)x^2}dx . \tag{4.3.13}$$

The latter integral is of the type:

$$f(t)=\int_0^\infty \sin\,(t(ax^2+bx))e^{-\frac{1}{2}(t^2\sigma_\theta^2+1)x^2}dx\ .$$

If $f(t)=0$ everywhere in t then also:

$$0=f'(0)=\int_0^\infty (ax^2+bx)e^{-\frac{1}{2}x^2}dx=a\frac{\sqrt{\pi}}{\sqrt{2}}+b\quad ,$$

$$0=f'''(0)=-\int_0^\infty (ax^2+bx)^3e^{-\frac{1}{2}x^2}dx=-a^3\int_0^\infty x^6e^{-\frac{1}{2}x^2}dx-3a^2b\int_0^\infty x^5e^{-\frac{1}{2}x^2}dx$$

$$-3ab^2\int_0^\infty x^4e^{-\frac{1}{2}x^2}dx-b^3\int_0^\infty x^3e^{-\frac{1}{2}x^2}dx\ .$$

Substituting $b=-a\frac{\sqrt{\pi}}{\sqrt{2}}$ in the last equality we obtain

$$0=a^3\{-\int_0^\infty x^6e^{-\frac{1}{2}x^2}dx+3\frac{\sqrt{\pi}}{\sqrt{2}}\int_0^\infty x^5e^{-\frac{1}{2}x^2}-3\frac{\pi}{2}\int_0^\infty x^4e^{-\frac{1}{2}x^2}+\frac{\pi^{3/2}}{2^{3/2}}\int_0^\infty x^3e^{-\frac{1}{2}x^2}dx\}\quad ,$$

hence $a=b=0$.

So we see now that $f(t)=0$ everywhere in t if and only if $a=b=0$. Consequently, $s(t,\theta)=0$ everywhere in t if and only if $\theta_1=\theta_1^0$ and $\theta_2=\theta_2^0$. Thus assumption 4.3.2 is satisfied.

Next we show for the case under review that $\Sigma_2(\theta_0)$ is nonsingular. We substitute $w(\underline{z}_j)r(\underline{z}_j,\theta)$ in (4.2.13):

$$\underline{c}_{n,1}(t,\theta_0)=-\frac{1}{n}\sum_{j=1}^n t\,\cos(t.w(\underline{z}_j)r(\underline{z}_j,\theta_0))w(\underline{z}_j)\underline{y}_{2,j}=$$

$$=-\frac{1}{n}\sum_{j=1}^n t\,\cos(t.x_{1,j}\underline{u}_{1,j})x_{1,j}\{\frac{\alpha\theta_2^0x_{1,j}+\beta x_{2,j}}{1-\alpha\theta_1^0}+\frac{\alpha\underline{u}_{1,j}+\underline{u}_{2,j}}{1-\alpha\theta_1^0}\}\ ,\qquad (4.3.14)$$

$$\underline{c}_{n,2}(t,\theta_0)=-\frac{1}{n}\sum_{j=1}^n t\,\cos(t.w(\underline{z}_j)r(\underline{z}_j,\theta_0))w(\underline{z}_j)x_{1,j}$$

$$=-\frac{1}{n}\sum_{j=1}^n t\,\cos(t.x_{1,j}\underline{u}_{1,j})x_{1,j}^2\ .\qquad (4.3.15)$$

Since $E\,\cos(t\underline{u}_{1,j})\underline{u}_{1,j}=0$ and $E\,\cos(t\underline{u}_{1,j})\underline{u}_{2,j}=E\,\cos(t\underline{u}_{1,j})E\underline{u}_{2,j}=0$ we obtain:

$$E\underline{c}_{n,2}(t,\theta_0)=-\frac{1}{n}\sum_{j=1}^n tE\,\cos(tx_{1,j}\underline{u}_{1,j})x_{1,j}^2=-\frac{1}{n}\sum_{j=1}^n tx_{1,j}^2e^{-\frac{1}{2}t^2x_{1,j}^2}\quad ,\qquad (4.3.16)$$

$$E\underline{c}_{n,1}(t,\theta_0)=+\frac{\alpha\theta_2^0}{1-\alpha\theta_1^0}E\underline{c}_{n,2}(t,\theta_0)-\frac{\beta}{1-\alpha\theta_1^0}\frac{1}{n}\sum_{j=1}^{n}tx_{1,j}x_{2,j}e^{-\frac{1}{2}t^2x_{1,j}^2}\ . \qquad (4.3.17)$$

Furthermore, from theorem 2.2.15 and (4.3.10) through (4.3.12) it follows:

$$c_2(t,\theta_0)=\lim E\underline{c}_{n,2}(t,\theta_0)=-t\int x^2e^{-\frac{1}{2}t^2x^2}dG_1(x)$$

$$=-t\frac{\sqrt{2}}{\sqrt{\pi}}\int_0^\infty x^2e^{-\frac{1}{2}(t^2+1)x^2}dx=\frac{-t}{(t^2+1)^{3/2}}\ , \qquad (4.3.18)$$

$$c_1(t,\theta_0)=\lim E\underline{c}_{n,1}(t,\theta_0)=\frac{-\alpha\theta_2^0}{1-\alpha\theta_1^0}\cdot\frac{t}{(t^2+1)^{3/2}}-\frac{\beta}{1-\alpha\theta_1^0}t\iint x_1x_2e^{-\frac{1}{2}t^2x_1^2}dG_1(x_1)dG_2(x_2)$$

$$=\frac{-\alpha\theta_2^0}{1-\alpha\theta_1^0}\cdot\frac{t}{(t^2+1)^{3/2}}-\frac{\beta}{1-\alpha\theta_1^0}\frac{\sqrt{2}}{\sqrt{\pi}}t\int_0^\infty xe^{-\frac{1}{2}(t^2+1)x^2}dx$$

$$=\frac{-\alpha\theta_2^0}{1-\alpha\theta_1^0}\cdot\frac{t}{(t^2+1)^{3/2}}-\frac{\sqrt{2}}{\sqrt{\pi}}\cdot\frac{\beta}{1-\alpha\theta_1^0}\cdot\frac{t}{t^2+1}\ , \qquad (4.3.19)$$

provided that $\sup_n \frac{1}{n}\sum_{j=1}^{n}|x_{2,j}|^{1+\delta}<\infty$ for some $\delta>0$. Now choose as weightfunction the density:

$$\phi(t)=\frac{6t}{(1+t^2)^4}\ ,\ t\geq 0, \qquad (4.3.20)$$

which obviously satisfies assumption 4.2.4. Then, using the equality

$$\int_0^\infty\frac{t^2}{(t^2+1)^\gamma}\frac{6t}{(1+t^2)^4}dt=3\left(\frac{1}{2+\gamma}-\frac{1}{3+\gamma}\right)$$

we obtain after some calculations

$$\Sigma_2(\theta_0)=3\begin{bmatrix}\left(\frac{\alpha\theta_2^0}{1-\alpha\theta_1^0}\right)^2\cdot\frac{1}{30}+\frac{2\sqrt{2}}{\sqrt{\pi}}\frac{\alpha\beta\theta_2^0}{(1-\alpha\theta_1^0)^2}\cdot\frac{4}{99}+\frac{2}{\pi}\left(\frac{\beta}{1-\alpha\theta_1^0}\right)^2\frac{1}{20}, & \frac{\alpha\theta_2^0}{1-\alpha\theta_1^0}\cdot\frac{1}{30}+\frac{\sqrt{2}}{\sqrt{\pi}}\frac{\beta}{1-\alpha\theta_1^0}\cdot\frac{4}{99} \\ \frac{\alpha\theta_2^0}{1-\alpha\theta_1^0}\cdot\frac{1}{30}+\frac{\sqrt{2}}{\sqrt{\pi}}\cdot\frac{\beta}{1-\alpha\theta_1^0}\cdot\frac{4}{99}\ , & \frac{1}{30}\end{bmatrix}$$

(4.3.21)

and thus

$$\det\Sigma_2(\theta_0) = \frac{1}{\pi}\left(\frac{\beta}{1-\alpha\theta_1^0}\right)^2 \cdot \frac{67}{108900} .$$

Note that if $\beta=0$ then not only $\det\Sigma_2(\theta_0)=0$, but also assumption 4.3.2 fails to hold, for (4.3.13) then becomes

$$s(t,\theta) = \frac{\sqrt{2}}{\sqrt{\pi}}\int_0^\infty \sin(t\{(\theta_2^0-\theta_2)+\frac{\alpha\theta_2^0}{1-\alpha\theta_1^0}(\theta_1^0-\theta_1)\}x^2)e^{-\frac{1}{2}(t^2\sigma_\theta^2+1)x^2}dx,$$

which is zero everywhere in t if $(\theta_2^0-\theta_2)+\frac{\alpha\theta_2^0}{1-\alpha\theta_1^0}(\theta_1^0-\theta_1)=0$. For every solution $\theta'=(\theta_1,\theta_2)$ of this equation the distribution function $H_\theta^*(u)$ from assumption 4.3.2 is therefore symmetric.

It is not very surprising that in the case $\beta=0$ our identification conditions fails to hold because then the system (4.3.4) also fails to satisfy the usual order condition for identification [see Theil (1971, theorem 10.2)]. This example shows that it is not always necessary to let the instrumental variable $\underline{w}_j$ be a function of all the exogenous variables of the system, and that there are cases where we may choose $\underline{w}_j$ to be a function of only the exogenous variables of the equation under review.

4.3.3 A scaling parameter and its impact on the asymptotic properties

The next point we want to discuss concerns the choice of the weightfunction $\phi(t)$. This point make sense because both the matrices $\Sigma_1(\theta_0)$ and $\Sigma_2(\theta_0)$ depend on $\phi(t)$, hence so does the efficiency of our minimum information estimator $\underline{\theta}_n^0$. First, observe from (4.2.25) and (4.2.34) that multiplying $\phi(t)$ with a constant does not affect $\Sigma_2(\theta_0)^{-1}\Sigma_1(\theta_0)\Sigma_2(\theta_0)^{-1}$; so without loss of generality we may assume that $\phi(t)$ is a density on $(0,\infty)$. Since for all $\gamma>0$, $\frac{1}{\gamma}\phi(\frac{t}{\gamma})$ is a density of the same class as $\phi(t)$, all the results of section 4.2 carry over when $\phi(t)$ is replaced by $\frac{1}{\gamma}\phi(\frac{t}{\gamma})$, $\gamma>0$. But in that case the asymptotic variance matrix $\Sigma_2(\theta_0)^{-1}\Sigma_1(\theta_0)\Sigma_2(\theta_0)^{-1}$ depends on γ, and so does $\mathrm{tr}[\Sigma_2(\theta_0)^{-1}\Sigma_1(\theta_0)\Sigma_2(\theta_0)^{-1}]$, which may be considered as a measure of the asymptotic efficiency of $\underline{\theta}_n^0$. The situation involved is thus comparable with that in the sections 3.2 and 3.3, where the asymptotic variance matrix of the robust M-estimator also depends on a scaling parameter γ.

In view of the above argument, we now replace $\phi(t)$ in (4.2.3) with $\frac{1}{\gamma}\phi(\frac{t}{\gamma})$:

$$\underline{S}_n^*(\theta|\gamma)=\int_0^\infty \underline{s}_n(t,\theta)^2 \frac{1}{\gamma}\,\phi(\frac{t}{\gamma})dt=\int_0^\infty \underline{s}_n(t\gamma,\theta)^2\phi(t)dt\ , \qquad (4.3.22)$$

where now also the instrumental variable $\underline{w}_j=w(\underline{z}_j)$ is taken into account:

$$\underline{s}_n(t,\theta)=\frac{1}{n}\sum_{j=1}^n \sin(t.w(\underline{z}_j)r(\underline{z}_j,\theta))\ . \qquad (4.3.23)$$

First it will be shown that for any compact interval Γ of $(0,\infty)$,

$$\sup_{(\theta,\gamma)\in\text{Ⓗ}\times\Gamma} |\underline{S}_n^*(\theta|\gamma)-S^*(\theta|\gamma)| \to 0 \text{ a.s.} \qquad (4.3.24)$$

where:

$$S^*(\theta|\gamma)=\int_0^\infty s(t\gamma,\theta)^2\ \phi(t)dt\ , \qquad (4.3.25)$$

with:

$$s(t,\theta)=\int \sin(t.w(z).r(z,\theta))dG(z). \qquad (4.3.26)$$

Doing so, we shall closely follow the proof of (4.2.7). Observe that from the assumptions 4.1.1, 4.2.2 and 4.3.1 and from theorem 2.3.3 it follows that (4.2.9) carries over for the present case[with (4.3.23) and (4.3.26)]. Moreover, from (4.2.10) it follows that

$$\sup_{(\theta,\gamma)\in\text{Ⓗ}\times\Gamma} |\underline{S}_n^*(\theta|\gamma)-S^*(\theta|\gamma)| \leq$$

$$\leq \sup_{\gamma\in\Gamma}\ \sup_{(t,\theta)\in T_\varepsilon\times\text{Ⓗ}} |\underline{s}_n(\gamma t,\theta)^2-s(\gamma t,\theta)^2|\int_0^\infty\phi(t)dt+\varepsilon$$

$$\leq \sup_{(t,\theta)\in T_\varepsilon^*\times\text{Ⓗ}} |\underline{s}_n(t,\theta)^2-s(t,\theta)^2|\int_0^\infty\phi(t)dt+\varepsilon\ , \qquad (4.3.27)$$

where $T_\varepsilon^*=[-M_\varepsilon.\sup_{\gamma\in\Gamma}\gamma,\ M_\varepsilon\sup_{\gamma\in\Gamma}\gamma]$. Thus (4.3.24) follows from (4.2.9), (4.3.27) and the fact that the supremum involved is a continuous function of the observation (see theorem 2.3.1) and hence a random variable.

Since the assumptions 4.3.2 and 4.2.4 imply that for every $\gamma>0$, θ_0 is a unique point in Ⓗ such that

$$0=S(\theta_0|\gamma)=\inf_{\theta\in\text{Ⓗ}} S(\theta|\gamma)\ , \qquad (4.3.28)$$

it follows now from (4.3.24) and lemma 3.4.1, similar as in theorem 3.4.1, that the random vector function $\underline{\theta}_n^0(\gamma)$ obtained from

$$\underline{S}^*_n(\underline{\theta}^0_n(\gamma)|\gamma) = \inf_{\theta\in\textcircled{H}} \underline{S}^*_n(\theta|\gamma) \ , \ \underline{\theta}^0_n(\gamma)\in\textcircled{H} \quad \text{a.s.} \ , \ \gamma > 0 \ , \qquad (4.3.29)$$

is pseudo-uniformly strongly consistent:

Theorem 4.3.1. Under the assumptions 4.1.1, 4.2.1, 4.2.2, 4.2.4, 4.3.1 and 4.3.2 it follows that $\underline{\theta}^0_n(\gamma)\to\theta_0$ a.s. uniformly on any compact subinterval Γ of $(0,\infty)$, provided that $\underline{\theta}^0_n(\gamma)$ is unique. Otherwise we only have pseudo-uniform convergence.

As already mentioned, the proof of theorem 4.2.2 for the present case is nearly the same as the original one. The changes are that $r(\underline{z}_j,\theta)$ in the assumptions 4.1.8 and 4.2.5 has to be replaced by $w(\underline{z}_j)r(\underline{z}_j,\theta)$ and that (4.2.26) has to be replaced by (4.3.3), and then all carries over. Thus we assume now

Assumption 4.3.3.
a) $\sup_n \frac{1}{n}\sum_{j=1}^n E\{\overline{w(\underline{z}_j)(\partial/\partial\theta_i)r(\underline{z}_j,\theta)}^{\textcircled{H}}\}^{1+\delta} < \infty$
and
b) $E\{\overline{w(\underline{z}_j)(\partial/\partial\theta_i)r(\underline{z}_j,\theta)}^{\textcircled{H}}\}^2 = O(j^\mu)$

for $i=1,2,\dots,q$ and some $\delta>0$, $\mu<1$, respectively,

and:

Assumption 4.3.4. $\sup_n \frac{1}{n}\sum_{j=1}^n E\{\overline{w(\underline{z}_j)(\partial/\partial\theta_{i_1})(\partial/\partial\theta_{i_2})r(\underline{z}_j,\theta)}^{\textcircled{H}}\}^{1+\delta} < \infty$

for $i_1,i_2=1,2,\dots,q$ and some $\delta>0$,

and then we conclude without further discussion

Theorem 4.3.2. Under the assumptions of theorem 4.3.1 and the additional assumptions 3.1.6, 4.3.3 and 4.3.4 we have

$$\sqrt{n}(\underline{\theta}^0_n(\gamma)-\theta_0)\to N_q[0,\Sigma_2(\theta_0|\gamma)^{-1}\Sigma_1(\theta_0|\gamma)\Sigma_2(\theta_0|\gamma)^{-1}] \text{ in distr.}$$

for any $\gamma>0$ for which $\Sigma_2(\theta_0|\gamma)$ is nonsingular, where

$$\Sigma_1(\theta|\gamma)=\left(\int_0^\infty\int_0^\infty \psi(\gamma t_1,\gamma t_2)c_{i_1}(\gamma t_1,\theta)c_{i_2}(\gamma t_2,\theta)\phi(t_1)\phi(t_2)dt_1dt_2\right), \qquad (4.3.30)$$

$$\Sigma_2(\theta|\gamma)=\left(\int_0^\infty c_{i_1}(\gamma t,\theta)c_{i_2}(\gamma t,\theta)\phi(t)dt\right), \qquad (4.3.31)$$

$$c_i(t,\theta)= -t\int\cos(t.h(z)r(z,\theta))w(z)(\partial/\partial\theta_i)r(z,\theta)dG(z) \qquad (4.3.32)$$

and $\psi(t_1,t_2)$ is defined by (4.3.3).

4.3.4 Estimation of the asymptotic variance matrix

Our next aim is to derive uniformly strongly consistent estimators of the matrices $\Sigma_1(\theta_0|\gamma)$ and $\Sigma_2(\theta_0|\gamma)$. Thus let:

$$\underline{c}_{n,i}(t,\theta) = -\frac{1}{n}\sum_{j=1}^{n} t\cos(t.w(\underline{z}_j).r(\underline{z}_j,\theta))w(\underline{z}_j)(\partial/\partial\theta_i)r(\underline{z}_j,\theta) \tag{4.3.33}$$

and

$$\underline{\psi}_n(t_1,t_2,\theta) = \frac{1}{n}\sum_{j=1}^{n}\sin(t_1 w(\underline{z}_j)r(\underline{z}_j,\theta))\sin(t_2.w(\underline{z}_j)r(\underline{z}_j,\theta)). \tag{4.3.34}$$

Since (4.2.15) carries over, it follows from lemma 3.3.1 that under the conditions of theorem 4.3.2,

$$\sup_{(t,\theta)\in T\times \textcircled{H}} |\underline{c}_{n,i_1}(t,\theta)\underline{c}_{n,i_2}(t,\theta) - c_{i_1}(t,\theta)c_{i_2}(t,\theta)| \to 0 \quad \text{a.s.} \tag{4.3.35}$$

for any compact interval T. Furthermore it follows from theorem 2.3.3 that for the same intervals T:

$$\sup_{(t_1,t_2,\theta)\in T\times T\times \textcircled{H}} |\underline{\psi}_n(t_1,t_2,\theta) - \int \sin(t_1 w(z)r(z,\theta))\sin(t_2 w(z)r(z,\theta))dG(z)| \to 0 \text{ a.s..} \tag{4.3.36}$$

From (4.3.35) we now have for any $\gamma_1 > 0$:

$$\sup_{t\in T}|\underline{c}_{n,i_1}(t,\underline{\theta}_n^0(\gamma_1))\underline{c}_{n,i_2}(t,\underline{\theta}_n^0(\gamma_1)) - c_{i_1}(t,\theta_0)c_{i_2}(t,\theta_0)| \to 0 \quad \text{a.s.} \tag{4.3.37}$$

and similar to (4.3.24) we thus conclude :

$$\sup_{\gamma\in\Gamma}\left|\int_0^\infty \underline{c}_{n,i_1}(\gamma t,\underline{\theta}_n^0(\gamma_1))\underline{c}_{n,i_2}(\gamma t,\underline{\theta}_n^0(\gamma_1))\phi(t)dt - \int_0^\infty c_{i_1}(\gamma t,\theta_0)c_{i_2}(\gamma t,\theta_0)\phi(t)dt\right| \to 0 \text{ a.s.} \tag{4.3.38}$$

for any $\gamma_1 > 0$ and any compact interval Γ of $(0,\infty)$.

In the same way we see from (4.3.36) and (4.3.37) that

$$\sup_{\gamma\in\Gamma}\left|\int_0^\infty\int_0^\infty \underline{\psi}_n(\gamma t_1, \gamma t_2,\underline{\theta}_n^0(\gamma_1))\underline{c}_{n,i_1}(\gamma t_1,\underline{\theta}_n^0(\gamma_1))\underline{c}_{n,i_2}(\gamma t_2,\underline{\theta}_n^0(\gamma_1))\phi(t_1)\phi(t_2)dt_1dt_2 - \int_0^\infty\int_0^\infty \psi(\gamma t_1, \gamma t_2)c_{i_1}(\gamma t_1,\theta_0)c_{i_2}(\gamma t_2,\theta_0)\phi(t_1)\phi(t_2)dt_1dt_2\right| \to 0 \quad \text{a.s.} \tag{4.3.39}$$

for any $\gamma_1 > 0$ and any compact interval Γ of $(0,\infty)$. Thus putting:

$$\underline{\Sigma}_{1,n}(\theta|\gamma) = \left(\int_0^\infty\int_0^\infty \underline{\psi}_n(\gamma t_1,\gamma t_2,\theta)\underline{c}_{n,i_1}(\gamma t_1,\theta)\underline{c}_{n,i_2}(\gamma t,\theta)\phi(t_1)\phi(t_2)dt_1dt_2\right), \tag{4.3.40}$$

$$\underline{\Sigma}_{2,n}(\theta|\gamma)=(\int_0^\infty \underline{c}_{n,i_1}(\gamma t,\theta)\underline{c}_{n,i_2}(\gamma t,\theta)\phi(t)dt) \,, \tag{4.3.41}$$

we now see, using theorem 2.3.2, that for any $\gamma_1>0$ and any compact interval Γ of $(0,\infty)$

$$\underline{\Sigma}_{1,n}(\underline{\theta}_n^0(\gamma_1)|\gamma) \to \Sigma_1(\theta_0|\gamma) \text{ a.s. uniformly on } \Gamma \tag{4.3.42}$$

$$\underline{\Sigma}_{2,n}(\underline{\theta}_n^0(\gamma_1)|\gamma) \to \Sigma_2(\theta_0|\gamma) \text{ a.s. uniformly on } \Gamma \,. \tag{4.3.43}$$

Using lemma 3.3.1 and the fact that the components of the inverse of a matrix are continuous functions of the components of this matrix, we now conclude from (4.3.42) and (4.3.43):

Theorem 4.3.3. Under the assumptions of theorem 4.3.2 we have for any $\gamma_1>0$

$$\underline{\Sigma}_{2,n}(\underline{\theta}_n^0(\gamma_1)|\gamma)^{-1}\underline{\Sigma}_{1,n}(\underline{\theta}_n^0(\gamma_1)|\gamma)\underline{\Sigma}_{2,n}(\underline{\theta}_n^0(\gamma_1)|\gamma)^{-1} \to \Sigma_2(\theta_0|\gamma)^{-1}\Sigma_1(\theta_0|\gamma)\Sigma_2(\theta_0|\gamma)^{-1}$$

a.s.

uniformly on any compact interval Γ of $(0,\infty)$ on which $\Sigma_2(\theta_0|\gamma)$ is non-singular.

4.3.5 A two-stage estimator

In this section we shall derive a more efficient two-stage minimum information estimation. This will be done in a similar way as in the case of weighted robust M-estimation. Thus using lemma 3.3.1 we conclude*) from theorem 4.3.3 that

$$\sup_{\gamma\in\Gamma}|\underline{T}_n^*(\gamma|\gamma_1)- T^*(\gamma)| \to 0 \quad \text{a.s.} \tag{4.3.44}$$

for any $\gamma_1>0$ and any compact interval Γ of $(0,\infty)$ on which $\Sigma_2(\theta_0|\gamma)$ is non-singular, where

$$\underline{T}_n^*(\gamma|\gamma_1)=\mathrm{tr}[\underline{\Sigma}_{2,n}(\underline{\theta}_n^0(\gamma_1)|\gamma)^{-1}\underline{\Sigma}_{1,n}(\underline{\theta}_n^0(\gamma_1)|\gamma)\underline{\Sigma}_{2,n}(\underline{\theta}_n^0(\gamma_1)|\gamma)^{-1}] \tag{4.3.45}$$

and

$$T^*(\gamma)=\mathrm{tr}[\Sigma_2(\theta_0|\gamma)^{-1}\Sigma_1(\theta_0|\gamma)\Sigma_2(\theta_0|\gamma)^{-1}] \,. \tag{4.3.46}$$

*) A direct conclusion is that $\underline{T}_n^*(\gamma|\gamma_1)\to T^*(\gamma)$ a.s. pseudo-uniformly in $\gamma\in\Gamma$. But it is not hard to verify that $|\underline{T}_n^*(\gamma|\gamma_1)- T^*(\gamma)|$ is a continuous function of γ and the observations $\underline{z}_1,\ldots,\underline{z}_n$, hence, using theorem 2.3.1, we conclude that $\sup_{\gamma\in\Gamma}|\underline{T}_n^*(\gamma|\gamma_1)- T^*(\gamma)|$ is a random variable.

Now let $\underline{\gamma}_n^0(\gamma_1)$ be a random variable satisfying

$$\underline{T}_n^*(\underline{\gamma}_n^0(\gamma_1)|\gamma_1) = \inf_{\gamma\in\Gamma} \underline{T}_n^*(\gamma|\gamma_1) \quad , \underline{\gamma}_n^0(\gamma_1)\in\Gamma \quad \text{a.s.} \tag{4.3.47}$$

and assume that

> <u>Assumption 4.3.5</u>. There is a unique point γ_0 in the compact interval Γ of $(0,\infty)$ such that $T^*(\gamma_0) = \inf_{\gamma\in\Gamma} T^*(\gamma)$.

Then it follows from lemma 3.1.3 that for any $\gamma_1 > 0$

$$\underline{\gamma}_n^0(\gamma_1) \to \gamma_0 \quad \text{a.s.} \tag{4.3.48}$$

Similar to the theorems 3.2.7 and 3.3.4 we now question whether

$$\sqrt{n}(\underline{\theta}_n^0(\underline{\gamma}_n^0(\gamma_1)) - \theta_0) \to N_q\left[0,\Sigma_2(\theta_0|\gamma_0)^{-1}\Sigma_1(\theta_0|\gamma_0)\Sigma_2(\theta_0|\gamma_0)^{-1}\right] \text{ in distr.. } \tag{4.3.49}$$

In view of the argument leading to these theorems it suffices to show that

$$\text{plim } \{\sqrt{n}(\partial/\partial\theta_i)\underline{S}_n^*(\theta_0|\underline{\gamma}_n^0(\gamma_1)) - \sqrt{n}(\partial/\partial\theta_i)\underline{S}_n^*(\theta_0|\gamma_0)\} = 0 \ . \tag{4.3.50}$$

Suppose that $\Gamma = \left[\alpha_1,\alpha_2\right]$. Since

$$\sqrt{n}(\partial/\partial\theta_i)\int_0^\infty \underline{s}_n(\gamma t,\theta_0)^2\phi(t)dt = 2\int_0^\infty \sqrt{n}\ \underline{s}_n(\gamma t,\theta_0)\underline{c}_{n,i}(\gamma t,\theta_0)\phi(t)dt$$

$$= 2.\frac{\gamma_0}{\gamma}\int_0^\infty \sqrt{n}\ \underline{s}_n(\gamma_0 t,\theta_0)\underline{c}_{n,i}(\gamma_0 t,\theta_0)\phi(\frac{\gamma_0}{\gamma}\ t)dt \tag{4.3.51}$$

it follows that

$$|\sqrt{n}(\partial/\partial\theta_i)\underline{S}_n^*(\theta_0|\gamma) - \sqrt{n}(\partial/\partial\theta_i)\underline{S}_n^*(\theta_0|\gamma_0)| \leq$$

$$\leq 2\int_0^\infty \sqrt{n}|\underline{s}_n(\gamma_0 t,\theta_0)||\underline{c}_{n,i}(\gamma_0 t,\theta_0)||\phi(\frac{\gamma_0}{\gamma}t) - \phi(t)|dt +$$

$$+2|\frac{\gamma_0}{\gamma} - 1|\int_0^\infty \sqrt{n}|\underline{s}_n(\gamma_0 t,\theta_0)||\underline{c}_{n,i}(\gamma_0 t,\theta_0)|\sup_{\gamma\in\left[\frac{\alpha_1}{\alpha_2},\frac{\alpha_2}{\alpha_1}\right]}\phi(\frac{t}{\gamma})dt \ . \tag{4.3.52}$$

Moreover, if ϕ is differentiable it follows from the mean value theorem that for some $\lambda\in\left[0,1\right]$

$$|\phi(\frac{\gamma_0}{\gamma}t)-\phi(t)|\leq|\frac{\gamma_0}{\gamma}-1||t\phi'(t.(1+\lambda(\frac{\gamma_0}{\gamma}-1)))|\leq|\frac{\gamma_0}{\gamma}-1|t.\sup_{\gamma\in\left[\frac{\alpha_1}{\alpha_2},\frac{\alpha_2}{\alpha_1}\right]}|\phi'(\frac{t}{\gamma})|. \quad (4.3.53)$$

Putting

$$\phi^*(t)=\sup_{\gamma\in\left[\frac{\alpha_1}{\alpha_2},\frac{\alpha_2}{\alpha_1}\right]}|\phi(\frac{t}{\gamma})|+t.\sup_{\gamma\in\left[\frac{\alpha_1}{\alpha_2},\frac{\alpha_2}{\alpha_1}\right]}|\phi'(\frac{t}{\gamma})|, \quad (4.3.54)$$

it follows then that

$$|\sqrt{n}(\partial/\partial\theta_i)\underline{S}_n^*(\theta_0|\gamma)-\sqrt{n}(\partial/\partial\theta_i)\underline{S}_n^*(\theta_0|\gamma_0)|\leq$$

$$2|\frac{\gamma_0}{\gamma}-1|\int_0^\infty\sqrt{n}|\underline{s}_n(\gamma_0t,\theta_0)||\underline{c}_{n,i}(\gamma_0t,\theta_0)|\phi^*(t)dt$$

$$\leq 2\gamma_0|\frac{\gamma_0}{\gamma}-1|\frac{1}{n}\sum_{j=1}^n|w(\underline{z}_j)r(\underline{z}_j,\theta_0)|\int_0^\infty\sqrt{n}|\underline{s}_n(\gamma_0t,\theta_0)|t\phi^*(t)dt. \quad (4.3.55)$$

Under the conditions of theorem 4.3.2 we have

$$\frac{1}{n}\sum_{j=1}^n|w(\underline{z}_j)r(\underline{z}_j,\theta_0)|\to\int|w(z)r(z,\theta_0)|dG(z)\quad\text{a.s.}, \quad (4.3.56)$$

while from (4.3.48)

$$\left|\frac{\gamma_0}{\underline{\gamma}_n(\gamma_1)}-1\right|\to 0\text{ a.s.}. \quad (4.3.57)$$

Moreover, if

> Assumption 4.3.6. $\phi(t)$ is chosen such that for any compact interval I of $(0,\infty)$, $t\sup_{\gamma\in I}|\phi(\frac{t}{\gamma})|$ and $t^2\sup_{\gamma\in I}|\phi'(\frac{t}{\gamma})|$ are integrable over $(0,\infty)$,

which is so if for example $\phi(t)$ is chosen to be a gamma density, then

$$E\int_0^\infty\sqrt{n}|\underline{s}_n(\gamma_0t,\theta_0)|t\phi^*(t)dt=\int_0^\infty E\sqrt{n}|\underline{s}_n(\gamma_0(t,\theta_0))|t\phi^*(t)dt$$

$$\leq\int_0^\infty\{En\,\underline{s}_n(\gamma_0t,\theta_0)^2\}^{\frac{1}{2}}t\phi^*(t)dt=\int_0^\infty\{\frac{1}{n}\sum_{j=1}^n E\sin(\gamma_0t\,w(\underline{z}_j)\underline{u}_j)^2\}^{\frac{1}{2}}t\phi^*(t)dt$$

$$\leq\int_0^\infty t\phi^*(t)dt<\infty, \quad (4.3.58)$$

hence the integral involved is a stochastically bounded random variable. From

(4.3.55) through (4.3.58) and lemma 4.2.1 it follows now that (4.3.50) holds and consequently that (4.3.49) holds. Thus:

Theorem 4.3.4. Under the assumptions of theorem 4.3.3 and the additional assumptions 4.3.5 and 4.3.6 we have for any $\gamma_1 > 0$

$\underline{\gamma}_n^0(\gamma_1) \rightarrow \gamma_0$ a.s. , $\underline{\theta}_n^0(\underline{\gamma}_n^0(\gamma_1)) \rightarrow \theta_0$ a.s.

and

$\sqrt{n}(\underline{\theta}_n^0(\underline{\gamma}_n^0(\gamma_1))-\theta_0) \rightarrow N_q[0, \Sigma_2(\theta_0|\gamma_0)^{-1}\Sigma_1(\theta_0|\gamma_0)\Sigma_2(\theta_0|\gamma_0)^{-1}]$

Remark: Note that the conclusion $\underline{\theta}_n^0(\underline{\gamma}_n^0(\gamma_1)) \rightarrow \theta_0$ follows from $\underline{\gamma}_n^0(\gamma_1) \rightarrow \gamma_0$ a.s., the pseudo-uniform strong consistency of $\underline{\theta}_n^0(\gamma)$ on Γ and from theorem 2.3.2.

4.3.6 Weak consistency

If only part a) of assumption 4.3.3 is satisfied then (4.3.42) and (4.3.43) only hold in pr. Consequently so does (4.3.44) and hence (4.3.48). Thus we have:

Theorem 4.3.5. If only part a) of assumption 4.3.3 is satisfied then "a.s." in the theorems 4.3.3 and 4.3.4 becomes "in pr.".

4.4 Miscellaneous notes on minimum information estimation

4.4.1 Remarks on the function $\underline{S}_n^*(\theta|\gamma)$

Applying the equality $\sin(a)\sin(b)=\frac{1}{2}\cos(a-b)-\frac{1}{2}\cos(a+b)$ to (4.3.22) and (4.3.23) and using the notation (4.3.1), we see that $\underline{S}_n^*(\theta|\gamma)$ can be written as

$$\underline{S}_n^*(\theta|\gamma)= \frac{1}{2n^2}\sum_{j_1=1}^n\sum_{j_2=1}^n \int_0^\infty \cos(\gamma t(r^*(\underline{z}_{j_1},\theta)-r^*(\underline{z}_{j_2},\theta)))\phi(t)dt$$

$$- \frac{1}{2n^2}\sum_{j_1=1}^n\sum_{j_2=1}^n \int_0^\infty \cos(\gamma t(r^*(\underline{z}_{j_1},\theta)+r^*(\underline{z}_{j_2},\theta)))\phi(t)dt . \qquad (4.4.1)$$

Furthermore, if $\phi(t)$ is a density on $(0,\infty)$ with characteristic function $\psi(\zeta)=\int_0^\infty e^{i\zeta t}\phi(t)dt$, then $\mathrm{Re}\psi(\zeta)=\int_0^\infty \cos(\zeta t)\phi(t)dt$ is the real part of ψ, hence

$$\underline{S}_n^*(\theta|\gamma)= \frac{1}{2n^2}\sum_{j_1=1}^n\sum_{j_2=1}^n \mathrm{Re}\psi(\gamma(r^*(\underline{z}_{j_1},\theta)-r^*(\underline{z}_{j_2},\theta)))$$

$$-\frac{1}{2n^2}\sum_{j_1=1}^n\sum_{j_2=1}^n \mathrm{Re}\psi(\gamma(r^*(\underline{z}_{j_1},\theta)+r^*(\underline{z}_{j_2},\theta))) =$$

$$= \frac{1}{2n^2}\sum_{j=1}^n(1-\mathrm{Re}\psi(2\gamma r^*(\underline{z}_j,\theta)) +$$

$$+ \frac{1}{n^2}\sum_{j_1=1}^{n-1}\sum_{j_2=j_1+1}^n \mathrm{Re}\psi(\gamma(r^*(\underline{z}_{j_1},\theta)-r^*(\underline{z}_{j_2},\theta)))$$

$$- \frac{1}{n^2}\sum_{j_1=1}^{n-1}\sum_{j_2=j_1+1}^n \mathrm{Re}\psi(\gamma(r^*(\underline{z}_{j_1},\theta)+r^*(\underline{z}_{j_2},\theta))) \qquad (4.4.2)$$

We thus see that for calculating the value of $\underline{S}_n^*(\theta|\gamma)$ for given θ and γ, the function $\mathrm{Re}\psi$ has to be calculated $n+2n(n-1)$ times, which may cost a lot of computer time when $\mathrm{Re}\psi$ is an expression involving functions such as cos(.), sin(.), exp (.) etc. But chosing $\phi(t)$ to be the gamma density

$$\phi(t)=t^{k-1}e^{-t}/\Gamma(k) \text{ ,} k=1,2,... \qquad (4.4.3)$$

we get

$$\mathrm{Re}\psi(\zeta)=\int_0^\infty \cos(\zeta t)\phi(t)dt=$$

$$= \tfrac{1}{2}\{\frac{1}{(1-i.\zeta)^k} + \frac{1}{(1+i.\zeta)^k}\} = \frac{(1+i.\zeta)^k+(1-i.\zeta)^k}{2(1+\zeta^2)^k} \quad , \qquad (4.4.4)$$

which is easily and hence cheaply calculated for small k. So (4.4.3) is recommended as weight function.

Now consider the linear errors in variables model $\underline{y}_1=\theta_1^0+ \theta_2^0\underline{x}_j+\underline{u}_j$, $\underline{d}_j=\underline{x}_j+\underline{v}_j$, with $(\theta_1^0,\theta_2^0)=(1,1)$, where the $\underline{x}_j$'s are unobservable. Put $\underline{z}_j'=(\underline{y}_j,\underline{d}_j)$, $r(\underline{z}_j,\theta)=\underline{y}_j-\theta_1-\theta_2\underline{d}_j$, $w(\underline{z}_j)=1$ and let the $\underline{x}_j$'s be random drawings from χ_2^2 times 5 and let the pairs $\begin{pmatrix}\underline{u}_j\\ \underline{v}_j\end{pmatrix}$ be random drawings from the bivariable normal distribution

$$N_2\left[\begin{pmatrix}0\\0\end{pmatrix},\begin{pmatrix}\sigma_u^2 & \rho\sigma_u\sigma_v\\ \rho\sigma_u\sigma_v & \sigma_v^2\end{pmatrix}\right]$$

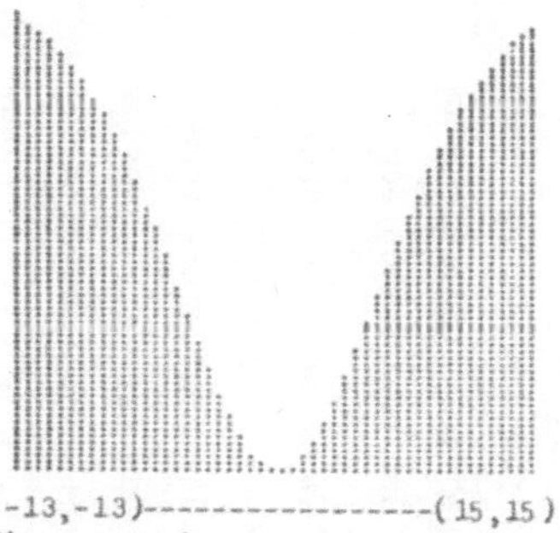

Figure 4.4.1: γ=0.01

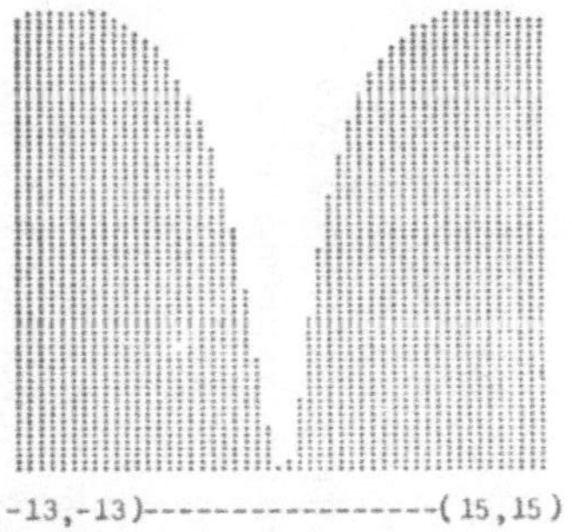

Figure 4.4.2: γ=0.1

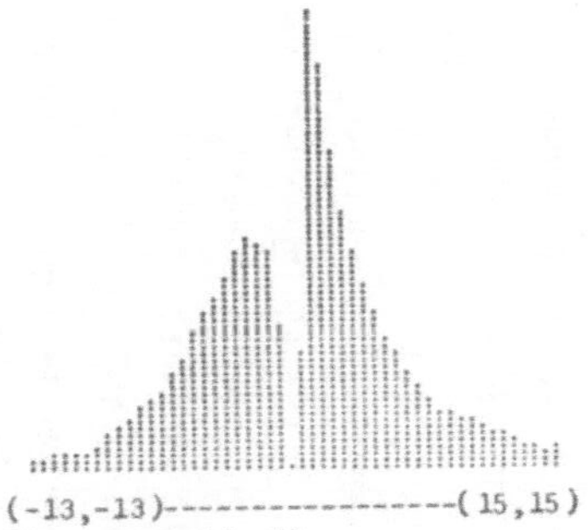

Figure 4.4.3: γ=1.

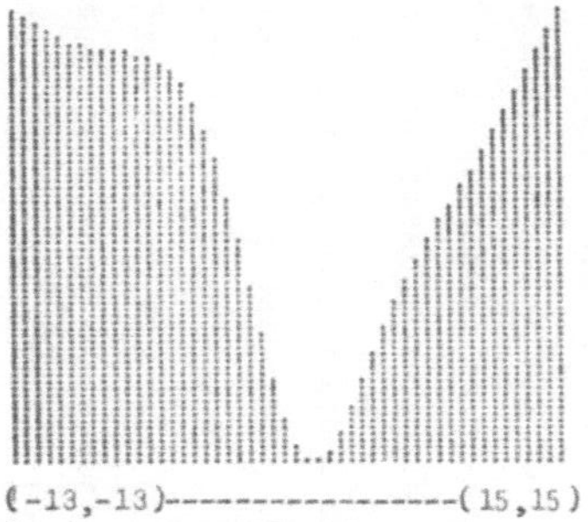

Figure 4.4.4: γ=0.01

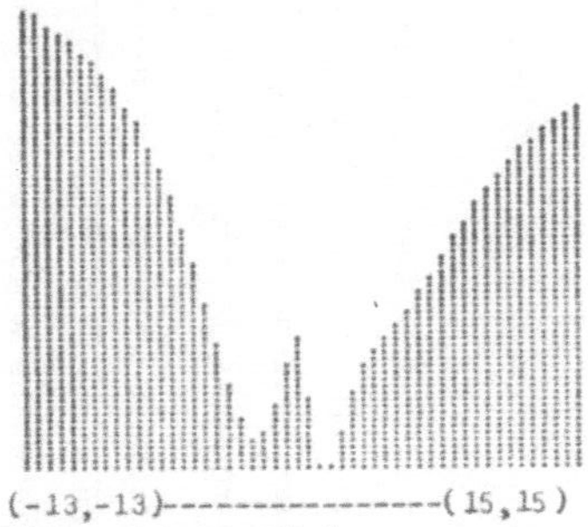

Figure 4.4.5: γ=0.1

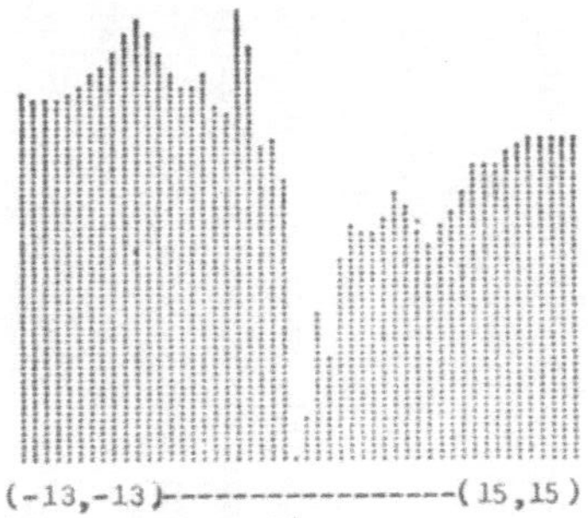

Figure 4.4.6: γ=1.

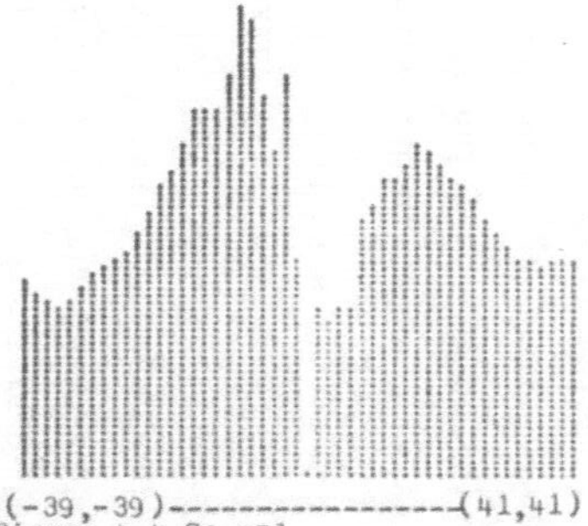

Figure 4.4.7: γ=1

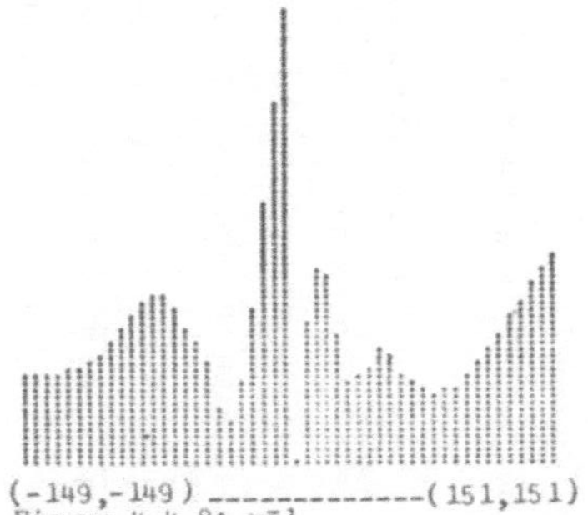

Figure 4.4.8: γ=1

where $\sigma_u^2=3$, $\rho_v^2=2$ and $\rho=0,8$. For this case we have plotted the function $\underline{S}_n^*(\theta|\gamma)$. The figures 4.4.1, 4.4.2 and 4.4.3 show the shape of $\underline{S}_n^*(\theta|\gamma)$ on the line between (-13, 13) and (15,15) for n=50 and $\gamma=0.01$, $\gamma=0.1$ and $\gamma=1$, respectively. Figure 4.4.3 shows beautifully a volcano shape of $\underline{S}_n^*(\theta|\gamma)$ with the true parameter point $\theta_0'=(1,1)$ on the bottom of the crater. By decreasing γ the crater of the volcano becomes wider, as is seen from the figures 4.4.1 and 4.4.2 .

One conclusion can already be drawn from these pictures. Minimizing $\underline{S}_n^*(\theta|\gamma)$ by an iterative procedure it is necessary to start with a point in (or on the wall of) the crater. For a realistic starting point this can always be provided by chosing γ small enough. Then carrying out the iteration, we get a point which lies on the (possibly flat) bottom of the crater. Using this point as a new starting point, we can increase now γ and continue the iteration. By theorem 4.3.1, all such "bottom" points are strongly consistent estimators of θ_0, provided that γ is chosen in a compact subset Γ of $(0,\infty)$.

The next example is the same errors in variables model, but now we have used an instrumental variable $\underline{w}_j$. This instrumental variable $\underline{w}_j$ is also drawn from a χ_2^2 distribution (times 5) but in order to ensure the dependence of $\underline{x}_j$ and $\underline{w}_j$ these two χ_2^2 distributions are derived from the marginal distributions of $N_2\left[\begin{pmatrix}0\\0\end{pmatrix}, \begin{pmatrix}1 & \lambda\\ \lambda & 1\end{pmatrix}\right]$ with some $\lambda \neq 0$. Again the figures 4.4.4, 4.4.5 and 4.4.6 show the shape of $\underline{S}_n^*(\theta|\gamma)$, n=50, on the line between (-13,-13) and (15,15) for $\gamma=0.01$, $\gamma=0.1$ and $\gamma=1.$, respectively. But we now see the appearance of local minima, due to the disturbance of $r(\underline{z}_j,\theta)$ with the instrumental variable $\underline{w}_j$. However, again we can get in the right crater by chosing γ small at the beginning of the iteration. Although this is not very clear from the figures, the basic shape is also a volcano, but with many secondary craters. This is seen from the figures 4.4.7 and 4.4.8 where for the case $\gamma=1$ the shape of the objective function on the lines between the points (-39,-39), (41,41) and the points (-149,-149), (151,151), respectively, is shown.

One of the pitfalls of nonlinear estimation is the difficulty to distinguish between local and global minima or maxima of the objective function. In our approach this is less problematic. We only have to vary the scaling factor γ to see whether or not a minimum is local or global, because the location of the global minimum will be much more invariant to shifts in this scaling factor than the locations of local minima.

4.4.2 A consistent initial value

Application of our method in the present stage is limited by computational difficulties, caused by the fact that we are dealing with double sums [compare (4.4.2)].But there is a way out. We can construct an approximation of the objective function (4.4.2) which generates consistent estimators and which is comparable with maximum likelihood estimation as far as computer time is concerned. The trick is to replace the n^2 pairs (j_1,j_2) of indices in the double sum (4.4.1) by a sample $\{(\underline{j}(1)_{n,1}, \underline{j}(2)_{n,1}) \ldots, (\underline{j}(1)_{n,n}, \underline{j}(2)_{n,n})\}$ from the distribution

$$P[(\underline{j}(1)_n, \underline{j}(2)_n) = (j_1,j_2)] = \frac{1}{n^2}, \quad j_1,j_2 = 1,2,\ldots,n, \tag{4.4.5}$$

so that the objective function (4.4.1) is now approximated by

$$\underline{S}_n^0(\theta|\gamma) = \frac{1}{2n}\sum_{i=1}^n \int_0^\infty \cos(\gamma t(r^*(\underline{z}_{\underline{j}(1)_{n,i}},\theta) - r^*(\underline{z}_{\underline{j}(2)_{n,i}},\theta)))\phi(t)dt$$

$$-\frac{1}{2n}\sum_{i=1}^n \int_0^\infty \cos(\gamma t(r^*(\underline{z}_{\underline{j}(1)_{n,i}},\theta) + r^*(\underline{z}_{\underline{j}(2)_{n,i}},\theta)))\phi(t)dt\,. \tag{4.4.6}$$

We shall prove that

Theorem 4.4.1. Under the conditions of theorem 4.3.1 and the assumptions 4.3.3 and 4.3.6,

$\sup_{(\theta,\gamma)\in \Theta \times \Gamma} |\underline{S}_n^*(\theta|\gamma) - \underline{S}_n^0(\theta|\gamma)| \to 0$ a.s. for any compact subinterval Γ of $(0,\infty)$,

which, similar to theorem 4.3.1, implies that also the random vector function $\underline{\tilde{\theta}}_n^0(\gamma)$ obtained from

$$\underline{S}_n^0(\underline{\tilde{\theta}}_n^0(\gamma)|\gamma) = \inf_{\theta\in\Theta} \underline{S}_n^0(\theta|\gamma), \quad \underline{\tilde{\theta}}_n^0(\gamma)\in\Theta \quad \text{a.s.} \tag{4.4.7}$$

is uniformly strongly consistent:

Theorem 4.4.2. Under the conditions of theorem 4.3.1 and the assumptions 4.3.3 and 4.3.6, $\underline{\tilde{\theta}}_n^0(\gamma)\to\theta_0$ a.s. unformly on any compact subinterval Γ of $(0,\infty)$.

Proof of theorem 4.4.1. Consider the following subset of Ⓗ x Γ:

$$N_\varepsilon(\theta_0,\gamma_0)=\{|\theta-\theta_0|<\varepsilon\}\times\{|\gamma-\gamma_0|<\varepsilon\}\cap Ⓗ\times\Gamma \tag{4.4.8}$$

where $(\theta_0,\gamma_0)\in$ Ⓗ x Γ. From the argument in the proof of theorem 2.3.3 we see that theorem 4.4.1 is proved if for any $\delta>0$ we can choose $\varepsilon>0$ so small that

$$\limsup_{n\to\infty}\ \sup_{(\theta,\gamma)\in N_\varepsilon(\theta_0,\gamma_0)}\{\underline{S}_n^0(\theta|\gamma)-\underline{S}_n^*(\theta|\gamma)\}\le\delta\quad\text{a.s.} \tag{4.4.9}$$

and

$$\liminf_{n\to\infty}\ \inf_{(\theta,\gamma)\in N_\varepsilon(\theta_0,\gamma_0)}\{\underline{S}_n^0(\theta|\gamma)-\underline{S}_n^*(\theta|\gamma)\}\ge-\delta\quad\text{a.s.} \tag{4.4.10}$$

for any $(\theta_0,\gamma_0)\in$ Ⓗ x Γ. Now put for convenience

$$\left.\begin{aligned}
&f_1(z_1,z_2,\theta,\gamma)=\int\cos(\gamma t(r^*(z_1,\theta)-r^*(z_2,\theta)))\phi(t)dt\ ,\\
&f_2(z_1,z_2,\theta,\gamma)=\int\cos(\gamma t(r^*(z_1,\theta)+r^*(z_2,\theta)))\phi(t)dt\ ,\\
&f(z_1,z_2,\theta,\gamma)=\tfrac{1}{2}f_1(z_1,z_2,\theta,\gamma)-\tfrac{1}{2}f_2(z_1,z_2,\theta,\gamma)\ ,\\
&N_\varepsilon=N_\varepsilon(\theta_0,\gamma_0)\ .
\end{aligned}\right\} \tag{4.4.11}$$

Then

$$\sup_{(\theta,\gamma)\in N_\varepsilon}\{\underline{S}_n^0(\theta|\gamma)-\underline{S}_n^*(\theta|\gamma)\}\le\sup_{(\theta,\gamma)\in N_\varepsilon}\underline{S}_n^0(\theta|\gamma)-\inf_{(\theta,\gamma)\in N_\varepsilon}\underline{S}_n^*(\theta|\gamma)\le$$

$$\le\frac{1}{n}\sum_{i=1}^n\sup_{(\theta,\gamma)\in N_\varepsilon}f(\underline{z}_{\underline{j}_{n,i}(1)},\underline{z}_{\underline{j}_{n,i}(2)},\theta,\gamma)-\frac{1}{n^2}\sum_{j_1=1}^n\sum_{j_2=1}^n\inf_{(\theta,\gamma)\in N_\varepsilon}f(\underline{z}_{j_1},\underline{z}_{j_2},\theta,\gamma)$$

$$\le\left|\frac{1}{n}\sum_{i=1}^n\sup_{(\theta,\gamma)\in N_\varepsilon}f(\underline{z}_{\underline{j}_{n,i}(1)},\underline{z}_{\underline{j}_{n,i}(2)},\theta,\gamma)-\frac{1}{n^2}\sum_{j_1=1}^n\sum_{j_2=1}^n\sup_{(\theta,\gamma)\in N_\varepsilon}f(\underline{z}_{j_1},\underline{z}_{j_2},\theta,\gamma)\right|$$

$$+\left|\frac{1}{n^2}\sum_{j_1=1}^n\sum_{j_2=1}^n\{\sup_{(\theta,\gamma)\in N_\varepsilon}f(\underline{z}_{j_1},\underline{z}_{j_2},\theta,\gamma)-\inf_{(\theta,\gamma)\in N_\varepsilon}f(\underline{z}_{j_1},\underline{z}_{j_2},\theta,\gamma)\}\right|$$

$$=\underline{x}_n+\underline{y}_n \tag{4.4.12}$$

say. Since $\sup_{(\theta,\gamma)\in N_\varepsilon}f(z_1,z_2,\theta,\gamma)$ is a uniformly bounded continuous (hence Borel measurable) function on $R^p\times R^p$ [compare theorem 2.3.1] it follows from lemma 4.4.1 below that

$$\underline{x}_n\to0\quad\text{a.s.} \tag{4.4.13}$$

Lemma 4.4.1. Let $f(z_1,z_2)$ be an uniformly bounded Borel measurable function on ZxZ, where Z is a subset of a Euclidean space. Let $(\underline{z}_j)$ be an arbitrary sequence of random vectors in Z.
Then

$$\frac{1}{n}\sum_{j=1}^{n} f(\underline{z}_{\underline{j}(1)_{n,i}}, \underline{z}_{\underline{j}(2)_{n,i}}) - \frac{1}{n^2}\sum_{j_1=1}^{n}\sum_{j_2=1}^{n} f(\underline{z}_{j_1}, \underline{z}_{j_2}) \to 0 \quad \text{a.s.} .$$

Proof. Put $\underline{x}_{n,i} = f(\underline{z}_{\underline{j}(1)_{n,i}}, \underline{z}_{\underline{j}(2)_{n,i}}) - \frac{1}{n^2}\sum_{j_1=1}^{n}\sum_{j_2=1}^{n} f(\underline{z}_{j_1}, \underline{z}_{j_2})$, $i=1,2,...$ Then the $\underline{x}_{n,i}$'s are uncorrelated for $i=1,2,...,n$, and moreover $E\{\underline{x}_{n,i_1}\cdot\underline{x}_{n,i_2}\cdot\underline{x}_{n,i_3}\cdot\underline{x}_{n,i_4}\}=0$ if one of the indices i_1,i_2,i_3,i_4 is unequal to each of the three others, which is so if (i_1,i_2,i_3,i_4) is not an element of the set

$$\{(i_1,i_2,i_3,i_4) \mid i_1=i_2=i_3=i_4 \text{ or } i_1=i_2\neq i_3=i_4 \text{ or } i_1=i_3\neq i_2=i_3 \text{ or } i_1=i_4\neq i_2=i_3\} .$$

Since this set has n+3n(n-1) elements, it follows now that

$$E\left(\frac{1}{n}\sum_{j=1}^{n}\underline{x}_{n,i}\right)^4 \leq \frac{n+3n(n-1)}{n^4}(2M)^4 < \frac{48M^4}{n^2} \text{ , where M is the uniform bound of } f(z_1,z_2).$$

Hence the lemma follows from Chebishev's inequality and the Borel-Cantelli lemma:

$$\sum_{n=1}^{\infty} P\left(\left|\frac{1}{n}\sum_{i=1}^{n}\underline{x}_{n,i}\right|>\varepsilon\right) \leq \sum_{n=1}^{\infty}\frac{48M^4}{\varepsilon^4 n^2} < \infty$$ □

We now continue the proof of theorem 4.4.1:
Observe from (4.4.11) and (4.3.53) that

$$|f_1(z_1,z_2,\theta,\gamma)-f_1(z_1,z_2,\theta,\gamma_0)| \leq$$

$$\leq \int_0^\infty \left|\frac{\gamma_0}{\gamma}\phi\left(\frac{t_0}{\gamma}\right) - \phi(t)\right|dt = \int_0^\infty \frac{|\gamma_0-\gamma|}{\gamma}\phi\left(\frac{t\gamma_0}{\gamma}\right)dt + \int_0^\infty \left|\phi\left(\frac{t\gamma_0}{\gamma}\right)-\phi(t)\right|dt$$

$$\leq \frac{|\gamma-\gamma_0|}{\alpha_1}\int_0^\infty \phi(t)dt + \frac{|\gamma-\gamma_0|}{\alpha_1}\int_0^\infty t \sup_{\gamma\in\left[\frac{\alpha_1}{\alpha_2},\frac{\alpha_2}{\alpha_1}\right]} \left|\phi'\left(\frac{t}{\gamma}\right)\right|dt$$

$$= |\gamma-\gamma_0|k_1 \tag{4.4.14}$$

where $\alpha_1 = \inf_{\gamma\in\Gamma}\gamma$, $\alpha_2 = \sup_{\gamma\in\Gamma}\gamma$ and by assumption 4.3.6, $k_1 < \infty$. Moreover from the mean value theorem we have:

$$|f_1(z_1,z_2,\theta,\gamma_0)-f_1(z_1,z_2,\theta_0,\gamma_0)|$$

$$\leq \gamma_0\{|r^*(z_1,\theta)-r^*(z_1,\theta_0)| + |r^*(z_2,\theta)- r^*(z_2,\theta_0)|\} \int_0^\infty t\phi(t)dt$$

$$\leq|\theta-\theta_0|\{\overline{|(\partial/\partial\theta)r^*(z_1,\theta)|}^{\text{(H)}} + \overline{|(\partial/\partial\theta)r^*(z_2,\theta)|}^{\text{(H)}}\}\alpha_2\int_0^\infty t\phi(t)dt. \tag{4.4.15}$$

Since the same applies for $f_2(z_1,z_2,\theta,\gamma)$, it follows from (4.4.11), (4.4.14) and (4.4.15) that

$$|f(z_1,z_2,\theta,\gamma)- f(z_1,z_2,\theta_0,\gamma_0)| \leq |\gamma-\gamma_0|k_1 +$$

$$|\theta-\theta_0|\{\overline{|(\partial/\partial\theta)r^*(z_1,\theta)|}^{\text{(H)}} + \overline{|(\partial/\partial\theta)r^*(z_2,\theta)|}^{\text{(H)}}\}k_2 , \tag{4.4.16}$$

where $k_2=\alpha_2\int_0^\infty t\phi(t)dt$. In its turn (4.4.16) implies

$$\frac{1}{n^2}\sum_{j_1=1}^n\sum_{j_2=1}^n \sup_{(\theta,\gamma)\in N_\varepsilon} f(\underline{z}_{j_1},\underline{z}_{j_2},\theta,\gamma)\leq \frac{1}{n^2}\sum_{j_1=1}^n\sum_{j_2=1}^n f(\underline{z}_{j_1},\underline{z}_{j_2},\theta_0,\gamma_0)$$

$$+\varepsilon\{k_1 + 2k_2 \frac{1}{n}\sum_{j=1}^n\overline{|(\partial/\partial\theta)r^*(\underline{z}_j,\theta)|}^{\text{(H)}}$$

and

$$\frac{1}{n^2}\sum_{j_1=1}^n\sum_{j_2=1}^n \inf_{(\theta,\gamma)\in N_\varepsilon} f(\underline{z}_{j_1},\underline{z}_{j_2},\theta,\gamma)\geq \frac{1}{n^2}\sum_{j_1=1}^n\sum_{j_2=1}^n f(\underline{z}_{j_1},\underline{z}_{j_2},\theta_0,\gamma_0)$$

$$-\varepsilon\{k_1+ 2k_2\frac{1}{n}\sum_{j=1}^n\overline{|(\partial/\partial\theta)r^*(\underline{z}_j,\theta)|}^{\text{(H)}}\} ;$$

hence

$$\underline{y}_n\leq 2\varepsilon\{k_1+ 2k_2 \frac{1}{n}\sum_{j=1}^n\overline{|(\partial/\partial\theta)r^*(\underline{z}_j,\theta)|}^{\text{(H)}}\}. \tag{4.4.17}$$

But from assumption 4.3.3 and theorem 2.3.3 it follows

$$\frac{1}{n}\sum_{j=1}^n\overline{|(\partial/\partial\theta)r^*(\underline{z}_j,\theta)|}^{\text{(H)}} \to \int\overline{|(\partial/\partial\theta)r^*(z,\theta)|}^{\text{(H)}} dG(z) \quad \text{a.s..} \tag{4.4.18}$$

Putting $k=2k_1+ 4k_2\int\overline{|(\partial/\partial\theta)r^*(z,\theta)|}^{\text{(H)}} dG(z)$ we thus have

$$\limsup_{n\to\infty}\underline{y}_n\leq \varepsilon k \quad \text{a.s. .} \tag{4.4.19}$$

Since now (4.4.9) follows from (4.4.12), (4.4.13) and (4.4.19) and since (4.4.10) can be proved in the same way, theorem 4.4.1 is proved by now, and so is theorem 4.4.2. □

Theorem 4.4.2 provides us with a good starting point for further iteration in order to find the global minimum of $\underline{S}_n^*(\theta|\gamma)$, but nothing is said about the asymptotic distribution of $\sqrt{n}(\tilde{\underline{\theta}}_n^0(\gamma)-\theta_0)$. Although it seems not impossible to derive this asymptotic distribution in the same way as for the original estimator $\underline{\theta}_n^0(\gamma)$, it will lead us to messy calculations and at least we need for this a generalisation of lemma 4.4.1 for unbounded functions. Besides, since $\tilde{\underline{\theta}}_n^0(\gamma)$ is based on less information than $\underline{\theta}_n^0(\gamma)$, we may expect that its performance in small samples as an estimator of θ_0 is less than that of $\underline{\theta}_n^0(\gamma)$. Thus we shall only use $\tilde{\underline{\theta}}_n^0$ as a starting point for further iterations. But now it is not necessary to find the global minimum of $\underline{S}_n^*(\theta|\gamma)$ because of the following theorem.

Theorem 4.4.3. Under the conditions of theorem 4.4.2, any random vector function $\tilde{\underline{\theta}}_n^*(\gamma)$ satisfying

$$\underline{S}_n^*(\tilde{\underline{\theta}}_n^*(\gamma)|\gamma) \leq \underline{S}_n^*(\tilde{\underline{\theta}}_n^0(\gamma)|\gamma) \;, \quad \tilde{\underline{\theta}}_n^*(\gamma) \in Ⓗ \quad \text{a.s.}$$

is pseudo-uniformly strongly consistent:

$$\tilde{\underline{\theta}}_n^*(\gamma) \to \theta_0 \quad \text{a.s. pseudo-uniformly on } \Gamma,$$

where Γ is a compact subinterval of $(0,\infty)$. If in addition $\underline{S}_n^*(\theta|\gamma)$ has a local minimum at $\tilde{\underline{\theta}}_n^*(\gamma)$ then $\tilde{\underline{\theta}}_n^*(\gamma)$ has the same limiting distribution as the estimator $\underline{\theta}_n^0(\gamma)$ considered in section 4.3.

This theorem can be proved by an easy further elaboration of lemma 3.4.1 and by noting that by lemma 3.1.5, $\sqrt{n}(\partial/\partial\theta_i)\underline{S}_n^*(\tilde{\underline{\theta}}_n^*(\gamma)|\gamma) \to 0$ a.s.

4.4.3 An upperbound of the variance matrix

Carrying out the minimization procedure (4.3.47) the computational difficulties are even much greater than for calculating the minimum of $\underline{S}_n^*(\theta|\gamma)$, especially because the elements of the matrix $\underline{\Sigma}_{1,n}(\theta_0|\gamma)$ are in fact triple-sums. However, since

Theorem 4.4.4. For any vector $x \in R^q$,

$$x'\Sigma_1(\theta_0|\gamma)x \leq x'\Sigma_2(\theta_0|\gamma)x.\int_0^\infty \phi(t)dt \tag{4.4.20}$$

and hence

$$x'\Sigma_2(\theta_0|\gamma)^{-1}\Sigma_1(\theta_0|\gamma)\Sigma_2(\theta_0|\gamma)^{-1}x \leq x'\Sigma_2(\theta_0|\gamma)^{-1}x\int_0^\infty \phi(t)dt \;, \tag{4.4.21}$$

the function $T^*(\gamma)$ defined in (4.3.46) is bounded from above by

$$\overline{T}^*(\gamma)=\int_0^\infty \phi(t)dt.tr(\Sigma_2(\theta_0|\gamma)^{-1}) \tag{4.4.22}$$

which is much easier to calculate. Thus if we choose $\underline{\gamma}_n^0(\gamma_1)\in\Gamma$ such that

$$\underline{\overline{T}}_n^*(\underline{\gamma}_n^0(\gamma_1)|\gamma_1)= \inf_{\gamma\in\Gamma} \underline{\overline{T}}_n^*(\gamma|\gamma_1) \tag{4.4.23}$$

where

$$\underline{\overline{T}}_n^*(\gamma|\gamma_1)=\int_0^\infty \phi(t)dt.tr(\underline{\Sigma}_{2,n}(\underline{\theta}_n^0(\gamma_1)|\gamma)^{-1}) \tag{4.4.24}$$

then under the conditions of theorem 4.3.4 the asymptotic variance matrix of $\sqrt{n}(\underline{\theta}_n^0(\underline{\gamma}_n^0(\gamma_1))-\theta_0)$ has a "minimum upperbound". Of course,the variance matrix involved is not necessarily minimal, but the above approach may be considered as the "second best" for the case where the procedure (4.3.47) is too expensive.

Proof of theorem 4.4.4. From (4.2.23), (4.3.3), (4.3.30), (4.3.31) and Schwarts inequality it follows that

$$x'\Sigma_1(\theta_0|\gamma)x=\int_0^\infty\int_0^\infty \psi(\gamma t_1,\gamma t_2)x'c(\gamma t_1,\theta_0)c(\gamma t_2,\theta_0)'x\phi(t_1)\phi(t_2)dt_1dt_2$$

$$\leq\sqrt{\int_0^\infty\int_0^\infty \psi(\gamma t_1,\gamma t_2)^2\phi(t_1)\phi(t_2)dt_1dt_2}\sqrt{\int_0^\infty\int_0^\infty \{x'c(\gamma t_1,\theta_0)\}^2\{x'c(\gamma t_2,\theta_0)\}^2\phi(t_1)\phi(t_2)dt_1dt_2}$$

$$\leq\int_0^\infty \phi(t)dt.\int_0^\infty \{x'c(\gamma t,\theta_0)\}^2\phi(t)dt=\int_0^\infty \phi(t)dt.x'\Sigma_2(\theta_0|\gamma)x$$

Now (4.4.21) follows from (4.4.20) by substituting $x=\Sigma_2(\theta_0|\gamma)^{-1}x$ in (4.4.20).

4.4.4 A note on the symmetry assumption

Our minimum information method is completely build up on the assumption that the error distribution is symmetric, hence if this is not the case this estimation method is useless. However, the minimum value of the objective function indicates when this occur, as we will show now.

First, if the error distribution is non symmetric then the limit function $S^*(\theta|\gamma)$ satisfies

$$\inf_{\theta\in(H)} S^*(\theta|\gamma) > 0 \quad \text{for every } \gamma > 0$$

because if there is a point $\theta_*\in(H)$ such that $S^*(\theta_*|\gamma)=0$ for some $\gamma>0$ then, since $\frac{1}{\gamma}\,\phi(\frac{t}{\gamma})$ is everywhere positive in $t \in (0,\infty)$, $s(\theta_*,t)=0$ everywhere in t and consequently we then could consider $r(\underline{z}_j,\theta_*)=\underline{u}_j^*$ as the errors (which may be heteroscedastic, but satisfying $\frac{1}{n}\sum_{j=1}^{n} P(\underline{u}_j^* \le u) \to H_*(u)$ properly, where $H_*(u)$ is a symmetric distribution). Thus if the symmetry assumption fails to hold then under (some of) the other assumption used in sections 4.3,

$$\text{plim}\; n^{\delta}\underline{S}_n^*(\underline{\theta}_n^0(\gamma)|\gamma) = \infty \text{ for all } \gamma>0 \text{ and all } \delta>0.$$

At the other hand, if the symmetry assumption is right then

$$E\; n\underline{S}_n^*(\theta_0|\gamma) = \int E\{\sqrt{n}.\underline{s}_n(\gamma t,\theta_0)\}^2 \phi(t)dt$$

$$= \frac{1}{n}\sum_{j=1}^{n}\int E[\sin(\gamma t\, \underline{u}_j w(\underline{z}_j))\}^2 \phi(t)dt \le \int \phi(t)dt$$

for all n, hence

$$\text{plim}\; n^{\delta}\underline{S}_n^*(\theta_0|\gamma)=0 \quad \text{for all } \gamma>0 \text{ and all } \delta<1$$

and consequently, since $\underline{S}_n^*(\theta_0|\gamma) \ge \underline{S}_n^*(\underline{\theta}_n^0(\gamma)|\gamma)$,

$$\text{plim}\; n^{\delta}\underline{S}_n^*(\underline{\theta}_n^0(\gamma)|\gamma)=0 \quad \text{for all } \gamma>0 \text{ and all } \delta<1.$$

So if we choose a priori numbers $\delta \in (0,1)$ and $M>0$ and if we decide that the error distribution is symmetric if $n^{\delta}\underline{S}_n^*(\underline{\theta}_n^0(\gamma)|\gamma) \le M$ and asymmetric if $n^{\delta}\underline{S}_n^*(\underline{\theta}_n^0(\gamma)|\gamma) > M$, we then have a consistent decision procedure for this assumption.

<u>Remark</u>: Since model (3.1.1) can be put in the form (4.1.2), we can also use this decision criterion for checking through the symmetry assumption in the robust M-estimation case.

5 NONLINEAR MODELS WITH LAGGED DEPENDENT VARIABLES

Dealing with models containing lagged dependent variables, econometricians are faced with the problem that in general the observations are dependently distributed. Consequently standard laws of large numbers and central limit theorems can no longer be used for deriving the asymptotic properties of the estimators involved. It is well known that for linear regression models this difficulty can be overcome by imposing some stability conditions on the model [see for example Malinvaud (1970, Ch. 14)]. In the nonlinear case, Sims (1976) has proved the strong consistency and asymptotic normality of a class of nonlinear robust M-estimators, including nonlinear least squares estimators, under the condition that the errors and the regressors are stationary and ergodic. However in our opinion the stationarity assumption is too restrictive.

The purpose of the first two sections of this chapter is to show that for a class of nonlinear evolutionary autoregressive stochastic processes various laws of large numbers and central limit theorems remain valid. These results will provide us with the tools necessary for generalizing the theory in the previous chapters 3 and 4 to non-stationary dynamic models, which will be done in section 5.3.

In section 5.1 we shall introduce a new stability concept. It will be shown that in the linear case this stability concept is closely related to the common stability condition of linear difference equations (a linear difference equation is stable if all the roots of the polynomial lag operator involved lie outside the unit circle, or with other words, if the roots of the characteristic equation involved have absolute values less than one). However, our concept carries over to some nonlinear autoregressive models.

5.1 Stochastic stability

5.1.1 Stochastically stable linear autoregressive processes

Roughly speaking, an autoregressive process is said to be stable if for every pair of initial values the differences between the two paths vanish as time tends to infinity. In the stochastic case these differences are random variables; hence stability of such stochastic processes should be described in terms of probabilities or mathematical expectations.

Let us consider the following first order linear autoregressive stochastic process

$$\underline{y}_t = \theta \underline{y}_{t-1} + \underline{u}_t \ , \ t=1,2,\ldots\ldots , \tag{5.1.1}$$

where $(\underline{u}_t)$ is a sequence of real valued random variables, the initial value $\underline{y}_0$ is a real valued random variable and θ is a given real number.
Clearly, repeated backwards substitution of (5.1.1) yields

$$\left.\begin{aligned} \underline{y}_t &= \theta^2 \underline{y}_{t-2} + \theta \underline{u}_{t-1} + \underline{u}_t \\ &= \ldots\ldots \\ &= \theta^m \underline{y}_{t-m} + \textstyle\sum_{j=0}^{m-1} \theta^j \underline{u}_{t-j} \qquad (1 \le m \le t) \\ &= \ldots\ldots \\ &= \theta^t \underline{y}_0 + \textstyle\sum_{j=0}^{t-1} \theta^j \underline{u}_{t-j} \end{aligned}\right\} \tag{5.1.2}$$

Put

$$\underline{y}_t^{(m)} = \begin{cases} \sum_{j=0}^{m-1} \theta^j \underline{u}_{t-j} & \text{if } 1 \le m \le t \\ \underline{y}_t & \text{if } m > t \end{cases} \tag{5.1.3}$$

We show now that if

$$|\theta| < 1 \tag{5.1.4}$$

and if

$$E|\underline{y}_0|^\alpha < \infty \ ; \sum_{t=0}^n E|\underline{u}_t|^\alpha = O(n^\gamma) \text{ for some } \alpha > 0 \text{ and some } \gamma \in R , \tag{5.1.5}$$

then there is a sequence (m_n) of positive integers satisfying

$$m_n = o(n) \tag{5.1.6}$$

such that

$$\lim_{n\to\infty} \frac{1}{n} \sum_{t=1}^n P(|\underline{y}_t - \underline{y}_t^{(m_n)}| > \delta) = 0 \text{ for every } \delta > 0 \tag{5.1.7}$$

This result will be the core of our new stability concept.

For showing (5.1.7) we observe from (5.1.2) and (5.1.3) that for $m \le t$

$$\begin{aligned} P(|\underline{y}_t - \underline{y}_t^{(m)}| > \delta) &= P(|\underline{y}_{t-m}| > \frac{\delta}{|\theta|^m}) = \\ &= P(|\theta^{t-m} \underline{y}_0 + \textstyle\sum_{j=0}^{t-m-1} \theta^j \underline{u}_{t-m-j}| > \frac{\delta}{|\theta|^m}) \\ &\le \frac{E|\theta^{t-m} \underline{y}_0 + \sum_{j=0}^{t-m-1} \theta^j \underline{u}_{t-m-j}|^\alpha}{(\frac{\delta}{|\theta|^m})^\alpha} \\ &\le \frac{2^\alpha |\theta|^{\alpha(t-m)} E|\underline{y}_0|^\alpha + 2^\alpha (t-m)^\alpha \sum_{j=0}^{t-m-1} |\theta|^{\alpha . j} E|\underline{u}_{t-m-j}|^\alpha}{(\frac{\delta}{|\theta|^m})^\alpha} , \end{aligned} \tag{5.1.8}$$

where the inequalities involved follows from Chebishev's inequality and twice inequality (2.1.30). Thus from (5.1.4), (5.1.5) and (5.1.8), we have

$$\frac{1}{n}\sum_{j=1}^{n}P(|\underline{y}_t-\underline{y}_t^{(m_n)}|>\delta)\leq(\frac{2}{\delta})^{\alpha}(\frac{1}{n}\sum_{t=m_n}^{n}|\theta|^{\alpha t})E|\underline{y}_0|^{\alpha}+(\frac{2}{\delta})^{\alpha}|\theta|^{\alpha m_n}\, n^{\alpha}\sum_{t=1}^{n}E|\underline{u}_t|^{\alpha}\ , \tag{5.1.9}$$

hence

$$\frac{1}{n}\sum_{j=1}^{n}P(|\underline{y}_t-\underline{y}_t^{(m_n)}|>\delta)=O(n^{\alpha+\gamma}|\theta|^{\alpha m_n}). \tag{5.1.10}$$

Now choose for example

$$m_n=[n^{\varepsilon}] \text{ with } \varepsilon\in(0,1). \tag{5.1.11}$$

Because of $|\theta|<1$, $\lim_{n\to\infty} n^{\alpha+\gamma}|\theta|^{\alpha[n^{\varepsilon}]}=0$, hence (5.1.7) is proved by now.

This result gives rise to the following general definition.

Definition 5.1.1. Let $(\underline{y}_t)$ be a sequence of random vectors in a Euclidean space Y and let the $\underline{y}_t$'s for $t\geq 1$ have the structure:

$$\underline{y}_t=\phi_t(\underline{u}_t,\underline{u}_{t-1},\ldots,\underline{u}_1,\underline{y}_0,\underline{y}_{-1},\ldots,\underline{y}_{-k+1}) \tag{5.1.12}$$

say, where the $\underline{u}_t$'s are random vectors in a Euclidean space U and the ϕ_t's are Borel measurable mappings from $(\times_{j=1}^{t}U)\times(\times_{j=1}^{k}Y)$ into Y. If there is a sequence (m_n) of positive integers satisfying

$$m_n=o(n) \tag{5.1.13}$$

and if there is a double array $(\underline{y}_{t,n})$ of random vectors in Y with the structure

$$\underline{y}_{t,n}=\begin{cases}\phi_{t,n}(\underline{u}_t,\underline{u}_{t-1},\ldots,\underline{u}_{t-m_n+1}) & \text{if } 1\leq m_n\leq t\\ \underline{y}_t & \text{if } m_n>t\geq 1\end{cases} \tag{5.1.14}$$

such that for every $\delta>0$

$$\lim_{n\to\infty}\frac{1}{n}\sum_{t=1}^{n}P(|\underline{y}_t-\underline{y}_{t,n}|>\delta)=0, \tag{5.1.15}$$

where the $\phi_{t,n}$'s are Borel measurable mappings from $\times_{j=1}^{m_n}U$ into Y then the sequence $(\underline{y}_t)$ is said to be a stochastically stable process with respect to the sequence $(\underline{u}_t)$. The latter sequence will be called the base of $(\underline{y}_t)$. The sequence $(\underline{y}_{t,n})$ will be called the m_n-approximation of $(\underline{y}_t)$.

Remark: Observe that this definition does not require that either the sequence (m_n) or the m_n-approximation $(\underline{y}_{t,n})$ is unique. For example, in the case (5.1.1) an alternative m_n-approximation would be:

$$\underline{y}_{t,n}=\begin{cases}\theta^{m_n}E\underline{y}_{t-m_n}+\sum_{j=0}^{m_n-1}\theta^{j}\underline{u}_{t-j} & \text{if } 1\leq m_n\leq t\\ \underline{y}_t & \text{if } m_n\geq t,\end{cases}$$

provided that $E|\underline{y}_0|<\infty$, $E|\underline{u}_t|<\infty$ for $t\geq1$ and $\sum_{t=1}^{n}E|\underline{u}_t-E\underline{u}_t|=O(n^{\gamma})$ for some $\gamma\in R$. This also motivates the subscripts "t,n" of the mappings $\phi_{t,n}$.

We have just proved:

Theorem 5.1.1. Let $(\underline{u}_t)$ be a sequence of random variables such that
$\sum_{t=1}^{n} E|\underline{u}_t|^{\alpha}=O(n^{\gamma})$ for some $\alpha > 0$ and some $\gamma\in R$ (5.1.16)
Let $\underline{y}_0$ be a random variable such that $E|\underline{y}_0|^{\alpha} <\infty$, where α is the same as before. Define the random variables $\underline{y}_t$, $t\geq 1$, recursively by
$\underline{y}_t=\theta\underline{y}_{t-1}+\underline{u}_t$, $t=1,2,\ldots$,
where θ is a given real number satisfying $|\theta|<1$. Then $(\underline{y}_t)$ is a stochastically stable process with respect to the base $(\underline{u}_t)$.

This result is a special case of the following:

Theorem 5.1.2. Let the $\underline{u}_t$'s be as in theorem 5.1.1.
Let $\underline{y}_0,\underline{y}_{-1},\ldots,\underline{y}_{-k+1}$ be random variables such that
$E|\underline{y}_{-j}|^{\alpha}< \infty$ for $j=0,1,\ldots,k-1$. (5.1.17)
Define the random variables $\underline{y}_t$, $t \geq 1$, recursively by
$\underline{y}_t=\theta_1\underline{y}_{t-1}+\ldots+ \theta_k\underline{y}_{t-1} + \underline{u}_t$, $t=1,2,\ldots$, (5.1.18)
where the θ_j's are given real variables. If all the roots of the equation
$\lambda^k=\theta_1 \lambda^{k-1}+ \theta_2\lambda^{k-2}+\ldots+ \theta_k$ (5.1.19)
have absolute values less than 1, then $(\underline{y}_t)$ is stochastically stable with respect to the base $(\underline{u}_t)$.

Before we prove theorem 5.1.2, we shall draw some further conclusions from the proof of the theorem 5.1.1. Thus from (5.1.9) we see that if we choose m_n proportional to $[n^{\varepsilon}]$, where ε is arbitrarily chosen in (0,1), then under the conditions of theorem 5.1.1

$$\lim_{n\to\infty} n^{\beta}\sum_{t=1}^{n}P(|\underline{y}_t- \underline{y}_t^{(m_n)}|> \frac{\delta}{n^{\rho}})=0 \tag{5.1.20}$$

for any $\beta\in R$ and $\rho\in R$. Moreover, from (5.1.3) it follows that the random variables $\underline{y}_t^{(m)}$ involved have for $1\leq m\leq t$ the structure

$$\underline{y}_t^{(m)}=\sum_{j=0}^{m-1} c_{t,m,j}\, \underline{u}_{t-j} , \tag{5.1.21}$$

where the $c_{t,m,j}$'s*) are constants such that

$$\sup_{t\geq 1}\ \max_{1\leq m\leq t}\ \sum_{j=0}^{m-1}|c_{t,m,j}|< \infty . \tag{5.1.22}$$

These results, together with the following lemma, will be used for proving theorem 5.1.2.

---

*) In fact $c_{t,m,j}=\theta^j$ with $|\theta|<1$, but the more general representation (5.1.21) will be convenient later on in the proof of theorem 5.1.2.

Lemma 5.1.1. Let $(\underline{u}_t)$, $(\underline{y}_t)$ and $(\underline{z}_t)$ be sequences of random variables. Let $(m_n^{(1)})$ and $(m_n^{(2)})$ be sequences of positive integers, both proportional to $[n^\varepsilon]$, where ε is some number in $(0,1)$. Thus $m_n^{(1)}=k_1[n^\varepsilon]$ and $m_n^{(2)}=k_2[n^\varepsilon]$, where k_1 and k_2 are given positive integer constants. Let

$$\underline{y}_t^{(m)}=\begin{cases}\sum_{j=0}^{m-1} c_{t,m,j}\,\underline{u}_{t-j} & \text{if } 1\le m\le t\\ \underline{y}_t & \text{if } m>t\geq 1,\end{cases} \tag{5.1.23}$$

$$\underline{z}_t^{(m)}=\begin{cases}\sum_{j=0}^{m-1} d_{t,m,j}\,\underline{y}_{t-j} & \text{if } 1\le m\le t\\ \underline{z}_t & \text{if } m>t\geq 1,\end{cases} \tag{5.1.24}$$

where the $c_{t,m,j}$'s and $d_{t,m,j}$'s satisfy

$$\sup_{t\geq 1}\max_{1\le m\le t}\sum_{j=0}^{m-1}|c_{t,m,j}|=M_1<\infty, \tag{5.1.25}$$

$$\sup_{t\geq 1}\max_{1\le m\le t}\sum_{j=0}^{m-1}|d_{t,m,j}|=M_2<\infty\ . \tag{5.1.26}$$

Suppose that for <u>any</u> $\beta\in R$, <u>any</u> $\rho\in R$, and <u>any</u> $\delta>0$

$$\lim_{n\to\infty} n^\beta\sum_{t=1}^{n}P(|\underline{y}_t-\underline{y}_t^{(m_n^{(1)})}|>\frac{\delta}{n^\rho})=0\ , \tag{5.1.27}$$

$$\lim_{n\to\infty} n^\beta\sum_{t=1}^{n}P(|\underline{z}_t-\underline{z}_t^{(m_n^{(2)})}|>\frac{\delta}{n^\rho})=0\ . \tag{5.1.28}$$

Then there exist random variables $\underline{z}_t^{*(m)}$ with the structure

$$\underline{z}_t^{*(m)}=\begin{cases}\sum_{j=0}^{m-1} e_{t,m,j}\,\underline{u}_{t-j} & \text{if } 1\le m\le t\\ \underline{z}_t & \text{if } m>t\geq 1,\end{cases} \tag{5.1.29}$$

where

$$\sup_{t\geq 1}\max_{1\le m\le t}\sum_{j=0}^{m-1}|e_{t,m,j}|<\infty \tag{5.1.30}$$

such that for any $\beta\in R$, any $\rho\in R$ and any $\delta>0$,

$$\lim_{n\to\infty} n^\beta\sum_{t=1}^{n}P(|\underline{z}_t-\underline{z}_t^{*(m_n^{(1)}+m_n^{(2)})}|>\frac{\delta}{n^\rho})=0$$

Proof: Let m be a positive integer. Let

$$m_1=\left[\frac{k_1}{k_1+k_2}\cdot m\right]\ ,\ m_2=m-m_1 .$$

Put

$$\underline{z}_t^{*(m)}=\begin{cases}\sum_{j=0}^{m_2-1} d_{t,m_2,j}\,\underline{y}_{t-j}^{(m_1)} & \text{if } 1\le m\le t\\ \underline{z}_t & \text{if } m>t\geq 1\ .\end{cases} \tag{5.1.31}$$

Substitution of (5.1.23) in (5.1.31) yields for $1\le m\le t$:

$$\underline{z}_t^{*(m)}=\sum_{j=0}^{m_2-1} d_{t,m_2,j}\sum_{\ell=0}^{m_1-1} c_{t-j,m_1,\ell}\underline{u}_{t-j-\ell}.$$

$$=\sum_{r=0}^{m-1}\{\sum_{\substack{0\le j\le m_2-1\\ 0\le \ell\le m_1-1\\ j+\ell=r}} d_{t,m_2,j}\, c_{t-j,m_1,\ell}\}\underline{u}_{t-r}$$

$$=\sum_{r=0}^{m-1} e_{t,m,r}\,\underline{u}_{t-r}\ ,$$

say. Moreover, from (5.1.25) and (5.1.26) it follows

$$\sum_{r=0}^{m-1}|c_{t,m,r}|\le \sum_{r=0}^{m-1}\sum_{\substack{0\le j\le m_2-1\\ 0\le \ell\le m_1-1\\ j+\ell=r}} |d_{t,m_2,j}||c_{t-j,m_2,\ell}|$$

$$=\sum_{j=0}^{m_2-1}\{|d_{t,m_2,j}|\sum_{\ell=0}^{m_1-1}|c_{t-j,m_2,\ell}|\}$$

$$\le \sum_{j=0}^{m_2-1}\{|d_{t,m_2,j}|\max_{1\le m_1\le t-j}\sum_{\ell=0}^{m_1-1}|c_{t-j,m_2,\ell}|\}$$

$$\le \sum_{j=0}^{m_2-1}|d_{t,m_2,j}|M_1$$

$$\le M_2.M_1<\infty\ .$$

Thus the $z_t^{*(m)}$'s defined by (5.1.31) have the structure (5.1.29) and (5.1.30).

Now, using (5.1.26) and the easy inequality

$$P(\sum_{j=1}^{K}|\underline{x}_j|>M)\le \sum_{j=1}^{K}P(|\underline{x}_j|>\frac{M}{K}) \tag{5.1.32}$$

we obtain for $1<m=m_1+m_2\le t\le n$,

$$P(|\underline{z}_t-\underline{z}_t^{*(m_1+m_2)}|>\frac{\delta}{n^\rho})\le$$

$$\le P(|\underline{z}_t-z_t^{(m_1)}|>\frac{\delta}{2n^\rho})+P(|\underline{z}_t^{(m_1)}-\underline{z}_t^{*(m_1+m_2)}|>\frac{\delta}{2n^\rho})$$

$$\le P(|\underline{z}_t-\underline{z}_t^{(m_1)}|>\frac{\delta}{2n^\rho})+P(\sum_{j=0}^{m_1-1}|\underline{y}_{t-j}-\underline{y}_{t-j}^{(m_2)}|>\frac{\delta}{2M_2n^\rho})$$

$$\le P(|\underline{z}_t-\underline{z}_t^{(m_1)}|>\frac{\delta}{2n^\rho})-\sum_{j=0}^{m_1-1}P(|\underline{y}_{t-j}-\underline{y}_{t-j}^{(m_2)}|>\frac{\delta}{2m_1M_2n^\rho})$$

$$\le P(|\underline{z}_t-\underline{z}_t^{(m_1)}|>\frac{\delta}{2n^\rho})+\sum_{t=1}^{n}P(|\underline{y}_t-y_t^{(m_2)}|>\frac{\delta}{2m_1M_2n^\rho}) \tag{5.1.33}$$

It is not hard to see now that the lemma follows from (5.1.27), (5.1.28) and (5.1.33). □

Proof of theorem 5.1.2:

Put

$$\phi(L)=1-\theta_1 L-\ldots-\theta_k L^k \ , \tag{5.1.34}$$

where L is the lag operator $L^j \underline{y}_t=\underline{y}_{t-j}$. Then model (5.1.18) becomes

$$\phi(L)\underline{y}_t=\underline{u}_t \ , \ t=1,2,\ldots\ldots \tag{5.1.35}$$

But $\phi(L)$ can be written as

$$\phi(L)=\prod_{j=1}^{k}(1-\lambda_j L) \ , \tag{5.1.36}$$

where the λ_j's are the roots of the equation (5.1.19). Let us for the moment assume that these λ_j's are all real valued. Put

$$\left.\begin{aligned} &\underline{z}_{0,t}=\prod_{j=1}^{k}(1-\lambda_j L)\underline{y}_t=\underline{u}_t \ , \\ &\underline{z}_{1,t}=\prod_{j=2}^{k}(1-\lambda_j L)\underline{y}_t \ , \\ &\underline{z}_{2,t}=\prod_{j=3}^{k}(1-\lambda_j L)\underline{y}_t \ , \\ &\quad\vdots \\ &\underline{z}_{k-1,t}=(1-\lambda_k L)\underline{y}_t \ , \\ &\underline{z}_{k,t}=\underline{y}_t \ . \end{aligned}\right\} \tag{5.1.37}$$

Then for j=1,2,...,k

$$\underline{z}_{j,t}=\lambda_j\underline{z}_{j,t}+\underline{z}_{j-1,t} \quad , \ t=1,2,\ldots \tag{5.1.38}$$

Thus the k-order autoregression (5.1.18) can be represented as a chain of k first-order regressions. This allows to prove theorem 5.1.2 by repeated application of lemma 5.1.1, as will be shown now.

From (5.1.20) through (5.1.22) it follows that if for j=1,2,...,k,

$$E|\underline{z}_{j,0}|^{\alpha}<\infty \ , \ \sum_{t=1}^{n}E|\underline{z}_{j-1,t}|^{\alpha}=O(n^{\gamma_j}) \tag{5.1.39}$$

for some $\alpha>0$ and some $\gamma_j\in R$, then for any $\beta\in R$, any $\rho\in R$, and any $\delta>0$

$$\lim_{n\to\infty} n^{\beta}\sum_{t=1}^{n}P(|\underline{z}_{j,t}-\underline{z}_{j,t}^{(m_n)}|>\frac{\delta}{n^{\rho}})=0 \ , \tag{5.1.40}$$

where $\underline{z}_{j,t}^{(m)}$ has the structure

$$\underline{z}_{j,t}^{(m)}=\begin{cases}\sum_{\ell=0}^{m-1}\lambda_j^{\ell}\,\underline{z}_{j-1,t-\ell}\ (=\sum_{\ell=0}^{m-1}d_{t,m,\ell}\,\underline{z}_{j-1,t-\ell}) & \text{if } 1\leq m\leq t \\ \underline{z}_{j,t} & \text{if } m>t\geq 1.\end{cases} \tag{5.1.41}$$

But then by repeated application of lemma 5.1.1 it follows that there is a sequence $(\underline{y}_t^{(m)})$ having the structure

$$\underline{y}_t^{(m)} = \begin{cases} \sum_{j=0}^{m-1} c_{t,m,j}\, \underline{u}_{t-j} & \text{if } 1 \le m \le t \\ \underline{y}_t & \text{if } m > t \end{cases} \tag{5.1.42}$$

with

$$\sup_{t\geq 1} \max_{1\le m\le t} \sum_{j=0}^{m-1} |c_{t,m,j}| < \infty , \tag{5.1.43}$$

such that for any $\beta \in R$, any $\rho \in R$ and any $\delta > 0$

$$\lim n^{\beta} \sum_{t=1}^{n} P(|\underline{y}_t - \underline{y}_t^{(m_n^*)}| > \frac{\delta}{n^{\rho}}) = 0 , \tag{5.1.44}$$

where (m_n^*) is a sequence of positive integers proportional to $[n^{\varepsilon}]$. Thus for the case that all the roots of equation (5.1.19) are real valued the theorem can now be proved by showing that (5.1.39) holds for 1=1,2,...,k. But for the case j=1 this is already implied by the conditions (5.1.16) and (5.1.17). Now, using inequality (2.1.13) it is easily shown that from (5.1.17) and (5.1.37) it follows

$$E|\underline{z}_{j,0}|^{\alpha} = E|\prod_{\ell=j+1}^{k} (1-\lambda_{\ell} L)\underline{y}_0|^{\alpha} < \infty \tag{5.1.45}$$

for j=1,2,...,k-1. Next, suppose that (5.1.39) holds for some j<k. Then, similar to (5.1.2), using inequality (2.1.13),

$$\sum_{t=1}^{n} E|\underline{z}_{j+1,t}|^{\alpha} = \sum_{t=1}^{n} E|\lambda_{j+1}^{t}\underline{z}_{j+1,0} + \sum_{\ell=0}^{t-1} \lambda_j^{\ell}\, \underline{z}_{j,t-\ell}|^{\alpha}$$

$$\le n\, 2^{\alpha}|\lambda_{j+1}|^{\alpha} E|\underline{z}_{j+1,0}|^{\alpha} + 2^{\alpha} n^{\alpha+1} \sum_{t=1}^{n} E|\underline{z}_{j,t}|^{\alpha}$$

$$= O(n^{1+\alpha+\gamma_j}) = O(n^{\gamma_{j+1}}) , \tag{5.1.46}$$

where $\gamma_{j+1} = 1+\alpha+\gamma_j$. Therefore, by induction, (5.1.39) holds for j=1,2,...,k; hence the theorem is proved for the case that all the λ_j's are real valued.

For the case that some of the λ_j's are complex valued the whole argument carries over, except that now the constants $c_{t,m,j}$'s in (5.1.42) and (5.1.43) may be complex, hence $\underline{y}_t^{(m)}$ may be a complex random variable for $1 \le m \le t$. But

$$P(|\underline{y}_t - \underline{y}_t^{(m_n)}| > \frac{\delta}{n^{\rho}}) = P((\underline{y}_t - \text{Re}\underline{y}_t^{(m_n)})^2 + (\text{Im}\, \underline{y}_t^{(m_n)})^2 > (\frac{\delta}{n^{\rho}})^2)$$

$$\geq P\{(\underline{y}_t - \text{Re}\underline{y}_t^{(m_n)})^2 > (\frac{\delta}{n^{\rho}})^2\} = P(|\underline{y}_t - \text{Re}\underline{y}_t^{(m_n)}| > \frac{\delta}{n^{\rho}}) \tag{5.1.47}$$

where

$$\text{Re}\underline{y}_t^{(m)} = \begin{cases} \sum_{j=0}^{m-1} (\text{Re}\, c_{t,m,j})\underline{u}_{t-j} & \text{if } 1 \le m \le t \\ \underline{y}_t & \text{if } m > t \geq 1. \end{cases} \tag{5.1.48}$$

Thus without loss of generality we may assume that the $c_{t,m,j}$'s are real valued. So theorem 5.1.2 is proved by now. □

The above argument also yields the following much stronger conclusion.

<u>Theorem 5.1.3</u>. Let the conditions of theorem 5.1.2 be satisfied. Let (m_n) be a sequence of positive integers proportional to $[n^{\varepsilon}]$, where ε is an arbitrarily chosen number in (0,1). Then there is an m_n-approximation $(\underline{y}_{t,n})$ such that for <u>any</u> $\beta \in R$, <u>any</u> $\rho \in R$ and <u>any</u> $\delta > 0$,
$\lim_{n\to\infty} n^{\beta} \sum_{t=1}^{n} P(|\underline{y}_t - \underline{y}_{t,n}| > \frac{\delta}{n^{\rho}}) = 0$.

5.1.2 <u>Multivariate stochastically stable processes</u>

Let $(\underline{y}_t)$ be a stochastically stable process with respect to the base $(\underline{u}_t)$, where $(\underline{u}_t)$ is a sequence of random vectors in a Euclidean space U. Let $(\underline{v}_t)$ be a sequence of random vectors in a Euclidean space V, and put

$\underline{w}_t' = (\underline{u}_t', \underline{v}_t')$

so that $(\underline{w}_t)$ is a sequence of random vectors in a Euclidean space W=UxV. Then it follows directly from definition 5.1.1 that $(\underline{y}_t)$ is also stochastically stable with respect to the base $(\underline{w}_t)$. Thus if $(\underline{z}_t)$ is a sequence of random vectors in a p-dimensional Euclidean space R^p such that for j=1,2,...,p the sequences $(\underline{z}_{j,t})$ are stochastically stable with respect to a base $(\underline{u}_{j,t})$, respectively, where $\underline{z}_{j,t}$ is the j-th component of $\underline{z}_t$ and $(\underline{u}_{j,t})$ is a sequence of randomvectors in a Euclidean space U_j, say then without loss of generality we may assume that each sequence $(\underline{z}_{j,t})$ is stochastically stable with respect to the common base $(\underline{u}_t)$, where

$\underline{u}_t' = (\underline{u}_{1,t}', \ldots, \underline{u}'_{p,t})$.

But such sequences $(\underline{z}_t)$ of random vectors are also stochastically stable with respect to the base $(\underline{u}_t)$:

<u>Theorem 5.1.3</u>. Let $(\underline{y}_t)$ and $(\underline{u}_t)$ be sequences of random vectors in the Euclidean spaces Y and U, respectively. Let $\underline{y}_{j,t}$ be the j-th component of $\underline{y}_t$. Then $(\underline{y}_t)$ is stochastically stable with respect to the base $(\underline{u}_t)$ if and only if for each j, $(\underline{y}_{j,t})$ is stochastically stable with respect to $(\underline{u}_t)$.

<u>Proof</u>: The "only if" part of the theorem follows directly from definition 5.1.1. Thus we assume now that for each j, $(\underline{y}_{j,t})$ is stochastically stable with respect

to $(\underline{u}_t)$. This means that for each j there is sequence $(m_{j,n})$ of positive integers satisfying $m_{j,n}=o(n)$ and an $m_{j,n}$-approximation $(\underline{y}_{j,t,n})$ with the structure

$$\underline{y}_{j,t,n}=\begin{cases}\varphi_{j,t,n}(\underline{u}_t,\underline{u}_{t-1},\ldots,\underline{u}_{t-m_{j,n}+1}) & \text{if } m_{j,n}\leq t\leq n\\ \underline{z}_{j,t} & \text{if } m_{j,n}>t\geq 1\end{cases} \tag{5.1.49}$$

such that for every $\delta>0$

$$\lim \tfrac{1}{n}\textstyle\sum_{j=1}^{n}P(|\underline{y}_{j,t}-\underline{y}_{j,t,n}|>\delta)=0\ . \tag{5.1.50}$$

Now put

$$\left.\begin{aligned} &m_n=\max_j m_{j,n}\ ,\\ &\underline{y}^*_{j,t,n}=\begin{cases}\varphi_{j,t,n}(\underline{u}_t,\underline{u}_{t-1},\ldots,\underline{u}_{t-m_{j,n}+1}) & \text{if } m_n\leq t\leq n\\ \underline{z}_{j,t} & \text{if } m_n>t\geq 1\ ,\end{cases}\\ &\underline{y}_{t,n}'=(\underline{y}^*_{1,t,n},\ldots,\underline{y}^*_{p,t,n})\ ,\end{aligned}\right\} \tag{5.1.51}$$

where p is the dimension of Y. Then

$$|\underline{y}^*_{j,t,n}-\underline{y}_{j,t,n}|=\begin{cases}0 & \text{if } t\geq m_n\\ |\underline{y}_{j,t}-\underline{y}_{j,t,n}| & \text{if } m_{j,n}\leq t<m_n\\ 0 & \text{if } t<m_{j,n}\ ,\end{cases} \tag{5.1.52}$$

hence, using inequality (5.1.32)

$$\begin{aligned}\tfrac{1}{n}\textstyle\sum_{t=1}^{n}P(|\underline{y}_t-\underline{y}_{t,n}|>\delta)&=\tfrac{1}{n}\textstyle\sum_{t=1}^{n}P(\sum_{j=1}^{p}|\underline{y}_{j,t}-\underline{y}^*_{j,t,n}|^2>\delta^2)\leq\\ &\leq \textstyle\sum_{j=1}^{p}\tfrac{1}{n}\sum_{t=1}^{n}P(|\underline{y}_{j,t}-\underline{y}^*_{j,t,n}|>\frac{\delta}{\sqrt{p}})\leq\\ &\leq \textstyle\sum_{j=1}^{p}\tfrac{1}{n}\sum_{t=1}^{n}P(|\underline{y}_{j,t}-\underline{y}_{j,t,n}|>\frac{\delta}{2\sqrt{p}})+\\ &+\textstyle\sum_{j=1}^{p}\tfrac{1}{n}\sum_{t=1}^{n}P(|\underline{y}_{j,t,n}-\underline{y}^*_{j,t,n}|>\frac{\delta}{2\sqrt{p}})\leq\\ &\leq 2\textstyle\sum_{j=1}^{p}\tfrac{1}{n}\sum_{t=1}^{n}P(|\underline{y}_{j,t}-\underline{y}_{j,t,n}|>\frac{\delta}{2\sqrt{p}})\ .\end{aligned} \tag{5.1.53}$$

Since $\underline{y}_{t,n}$ has obviously the structure (5.1.14) and $m_n=o(n)$, the theorem follows now from (5.1.50) and (5.1.53). □

5.1.3 Other examples of stochastically stable processes

We show now by some examples that the stochastic stability concept carries over for some nonlinear autoregressive stochastic processes. The first example is easy:

Example 1.

$$\underline{y}_t=\sin(\theta\underline{y}_{t-1})+\underline{u}_t \quad , t=1,2,... \tag{5.1.54}$$

where $(\underline{u}_t)$ is a sequence of random variables, $\underline{y}_0$ is a random variable and θ is a number satisfying $|\theta|<1$.

Define the sequence (ζ_t) recursively by

$$\zeta_t=\sin(\theta\zeta_{t-1}) \quad , t\geq 1 \; , \zeta_0=0 \; . \tag{5.1.55}$$

Then, similar to (5.1.2), using the mean value theorem

$$|\underline{y}_t-\zeta_t| \leq |\theta||\underline{y}_{t-1}-\zeta_{t-1}|+|\underline{u}_t| \leq$$

$$\cdots$$

$$\leq|\theta|^t|\underline{y}_0|+\textstyle\sum_{j=0}^{t-1}|\theta|^j|\underline{u}_{t-j}| \; . \tag{5.1.56}$$

Moreover, let $\underline{z}_{t-m}^{(m)}=\zeta_{t-m}$ a.s. and define for $t_*>t-m$

$$\underline{z}_{t_*}^{(m)}=\sin(\theta\underline{z}_{t_*-1}^{(m)})+\underline{u}_{t_*} \; . \tag{5.1.57}$$

Then similarly

$$|\underline{y}_t-\underline{z}_t^{(m)}|\leq|\theta|^m|\underline{y}_{t-m}-\underline{z}_{t-m}^{(m)}|=|\theta|^m|\underline{y}_{t-m}-\zeta_{t-m}| \; . \tag{5.1.58}$$

Combining (5.1.57) and (5.1.58) we obtain

$$|\underline{y}_t-\underline{z}_t^{(m)}|\leq|\theta|^t|\underline{y}_0|+|\theta|^m\textstyle\sum_{j=0}^{t-m-1}|\theta|^j|\underline{u}_{t-m-j}| \; . \tag{5.1.59}$$

But $\underline{z}_t^{(m)}$ has the structure

$$\underline{z}_t^{(m)}=\phi_{t,m}(\underline{u}_t,\underline{u}_{t-1},\ldots,\underline{u}_{t-m+1}) \; . \tag{5.1.60}$$

Thus if we put

$$\underline{y}_t^{(m)}=\begin{cases}\underline{z}_t^{(m)} & \text{if } t\geq m\\ \underline{y}_t & 1\leq t<m \quad ,\end{cases} \tag{5.1.61}$$

then it follows, similar to the proof of theorem 5.1.1, that if

$$E|\underline{y}_0|^\alpha<\infty \quad , \textstyle\sum_{t=1}^n E|\underline{u}_t|^\alpha=O(n^\gamma) \tag{5.1.62}$$

for some $\alpha>0$ and some γ, then $(\underline{y}_t)$ is stochastically stable with respect to the base $(\underline{u}_t)$.

This example is a special case of an autoregression of the type

$$\underline{y}_t=g(\underline{y}_{t-1})+\underline{u}_t, \; t=1,2,.....$$

where g is a contraction mapping [see Kolmogorov and Fomin (1957)], which means that there is a number $\delta\in(0,1)$ such that for every pair (y_1,y_2) in R x R,

$$|g(y_1)-g(y_2)|\leq\delta|y_1-y_2|.$$

It is easy to see that if in this case (5.1.62) is satisfied for some $\alpha>0$ en some γ, then also now (y_t) is stochastically stable with respect to (u_t).

Example 2.

$$\underline{y}_t=\prod_{j=1}^{k}\underline{y}_{t-j}^{\theta_j}\,e^{\underline{u}_t}\quad ,\ t=1,2,... \tag{5.1.63}$$

Example 3.

$$\underline{y}_t=\{\textstyle\sum_{j=1}^{k}\epsilon_j\underline{y}_{t-j}^{\rho}+\underline{u}_t\}^{\frac{1}{\rho}}\quad ,\ 0<\rho<1 \tag{5.1.64}$$

Both nonlinear autoregressions have the property that they can be made linear by appropriate transformation:

$$\underline{z}_t=\textstyle\sum_{j=1}^{k}\theta_j\,\underline{z}_{t-j}+\underline{u}_t\quad ,\ t=1,1,..... \tag{5.1.65}$$

where

$$\left.\begin{array}{l}\underline{z}_t=\log(\underline{y}_t)\ \text{ in the case of example 2 },\\ \\ \underline{z}_t=\underline{y}_t^{\rho}\ \text{ with } \rho\in(0,1) \text{ in the case of example 3 .}\end{array}\right\} \tag{5.1.66}$$

From theorem 5.1.2 it follows that if

$$\textstyle\sum_{j=0}^{n}E|\underline{u}_t|^{\alpha}=O(n^{\gamma}) \tag{5.1.67}$$

for some $\alpha>0$ and some γ, if

$$\left.\begin{array}{l}E|\log(\underline{y}_j)|^{\alpha}<\infty \text{ for } j=0,-1,...,-k+1 \text{ in the case of example 2}\\ E|\underline{y}_t|^{\rho\alpha}<\infty \text{ for } j=0,-1,...,-k+1 \text{ in the case of example 3}\end{array}\right\} \tag{5.1.68}$$

and if all the roots of the equation

$$\lambda^k=\theta_1\lambda^{k-1}+\ ...\ +\theta_k \tag{5.1.69}$$

have absolute values less than 1, then in both cases the sequence $(\underline{z}_t)$ is stochastically stable with respect to $(\underline{u}_t)$. But does this imply the stochastic stability of $(\underline{y}_t)$ with respect to $(\underline{u}_t)$? The answer is given by the following theorem.

Theorem 5.1.4. Let $(\underline{y}_t)$ be stochastically stable with respect to the base $(\underline{u}_t)$, where $(\underline{y}_t)$ is a sequence of random vectors in a Euclidean space Y Let $f(y)$ be a continuously differentiable real function on Y. Let G_t be the distribution function of $|\underline{y}_t|$. If $\frac{1}{n}\sum_{t=1}^{n}G_t$ is stochastically bounded*) then the sequence $(f(\underline{y}_t))$ is stochastically stable with respect to $(\underline{u}_t)$.

*) This means that for every $\varepsilon>0$ there exists an $M_\varepsilon>0$ such that

$$\int_{\{|y|>M_\varepsilon\}} d\left(\frac{1}{n}\textstyle\sum_{t=1}^{n}G_t(y)\right)\leq\varepsilon\quad \text{for all } n\geq 1.$$

Proof: Let $(\underline{y}_{t,n})$ be the m_n-approximation involved. From the mean value theorem it follows that for each t and n there is a random variable $\underline{\lambda}_{t,n}$ such that both

$$0 \leq \underline{\lambda}_{t,n} \leq 1 \quad \text{a.s.} \tag{5.1.70}$$

and

$$f(\underline{y}_t) - f(\underline{y}_{t,n}) = (y_t - \underline{y}_{t,n})(\partial/\partial y')f(\underline{y}_t + \underline{\lambda}_{t,n}(\underline{y}_{t,n} - \underline{y}_t)) . \tag{5.1.71}$$

Hence

$$|f(\underline{y}_t) - f(\underline{y}_{t,n})| \leq |\underline{y}_t - \underline{y}_{t,n}| \sup_{|x| \leq |\underline{y}_t - \underline{y}_{t,n}|} |(\partial/\partial y)f(\underline{y}_t + x)| \tag{5.1.72}$$

and consequently

$$P(|\underline{y}_t - \underline{y}_{t,n}| \leq \delta) \leq P(|f(\underline{y}_t) - f(\underline{y}_{t,n})| \leq \delta\psi_\delta(\underline{y}_t)) , \tag{5.1.73}$$

where

$$\psi_\delta(y) = \sup_{|x| \leq \delta} |(\partial/\partial y)f(y+x)| . \tag{5.1.74}$$

Since $\frac{1}{n}\sum_{t=1}^n G_t$ is stochastically bounded, it follows that for every $\varepsilon > 0$ there is an $M_\varepsilon > 0$ such that

$$\frac{1}{n}\sum_{t=1}^n P(|\underline{y}_t| \leq M_\varepsilon) > 1-\varepsilon \quad \text{for all } n \geq 1. \tag{5.1.75}$$

Moreover, from lemma 2.3.1 it follows that $\psi_\delta(y)$ is continuous on Y, hence

$$\sup_{|y| \leq M_\varepsilon} \psi_\delta(y) = K_{\varepsilon,\delta} < \infty . \tag{5.1.76}$$

Thus it follows from (5.1.73) through (5.1.76) and the stochastic stability of $(\underline{y}_t)$ that for any $\delta > 0$ and any $\varepsilon > 0$

$$1 = \lim_{n\to\infty} \frac{1}{n}\sum_{t=1}^n P(|\underline{y}_t - \underline{y}_{t,n}| \leq \delta)$$

$$\leq \limsup_{n\to\infty} \frac{1}{n}\sum_{t=1}^n P(|f(\underline{y}_t) - f(\underline{y}_{t,n})| \leq \delta\psi_\delta(\underline{y}_t))$$

$$\leq \limsup_{n\to\infty} \frac{1}{n}\sum_{t=1}^n P(|f(\underline{y}_t) - f(\underline{y}_{t,n})| \leq \delta\psi_\delta(y_t) \text{ and } |\underline{y}_t| \leq M_\varepsilon)$$

$$+ \limsup_{n\to\infty} \frac{1}{n}\sum_{t=1}^n P(|f(\underline{y}_t) - f(\underline{y}_{t,n})| \leq \delta\psi_\delta(\underline{y}_t) \text{ and } |\underline{y}_t| > M_\varepsilon)$$

$$\leq \limsup_{n\to\infty} \frac{1}{n}\sum_{t=1}^n P(|f(\underline{y}_t) - f(\underline{y}_{t,n})| \leq \delta K_{\varepsilon,\delta})$$

$$+ \limsup_{n\to\infty} \frac{1}{n}\sum_{t=1}^n P(|\underline{y}_t| > M_\varepsilon)$$

$$\leq \limsup_{n\to\infty} \frac{1}{n}\sum_{t=1}^n P(|f(\underline{y}_t) - f(\underline{y}_{t,n})| \leq \delta K_{\varepsilon,\delta}) + \varepsilon$$

Since for fixed ε(and M_ε), $K_{\varepsilon,\delta}$ is monotonic non-decreasing in $\delta > 0$, it follows that $\lim_{\delta\downarrow 0} \delta K_{\varepsilon,\delta} = 0$. Thus for arbitrarily chosen positive numbers δ^* and ε we can make δ

so small that $\delta K_{\varepsilon,\delta} \leq \delta^*$, hence

$$\limsup_{n\to\infty} \frac{1}{n}\sum_{t=1}^{n} P(|f(\underline{y}_t) - f(\underline{y}_{t,n})| \leq \delta^*) \geq 1-\varepsilon$$

for every $\delta^*>0$ and every $\varepsilon>0$. This implies that $\lim \frac{1}{n}\sum_{t=1}^{n} P(|f(\underline{y}_t) - f(\underline{y}_{t,n})|>\delta^*)=0$ for any $\delta^*>0$; and so the theorem is proved. □

We now return to the examples 2 and 3. Since the transformations $\underline{y}_t=e^{\underline{z}_t}$ and $\underline{y}_t=\underline{z}_t^{\frac{1}{\rho}}$ are continuously differentiable, it follows from the above theorem that the stochastic stability of $(\underline{z}_t)$ implies the stochastic stability of $(\underline{y}_t)$ if the distribution functions of the $|\underline{z}_t|$'s are stochastically bounded in mean. This is, for example, the case if the process $(\underline{z}_t)$ is stationary, or if the mean of the distribution functions of the $\underline{z}_t$'s converges properly.

If the function f(y) in theorem 5.1.4 has uniformly bounded continuous first partial derivatives then the condition that $\frac{1}{n}\sum_{j=1}^{n} G_t$ is stochastically bounded is not needed, as is easily verified from the proof. Thus

> Theorem 5.1.5. Let $(\underline{y}_t)$ be stochastically stable with respect to a base $(\underline{u}_t)$, where $(\underline{y}_t)$ is a sequence of random vectors in a Euclidean space Y. Let f(y) be a real function on Y with uniformly bounded continuous first partial derivatives. Then $(f(\underline{y}_t))$is stochastically stable with respect to $(\underline{u}_t)$.

The above list of examples of nonlinear stochastically stable processes is, of course,far from complete. We only wanted to show that stochastic stability may be a meaningful concept for nonlinear autoregressions. Comparing the result of theorem 5.1.3 with the definition 5.1.1 we obviously get the strong impression that stochastic stability is a very weak concept for linear autoregressions and that it therefore also is applicable for "nearly linear" nonlinear autoregressions.

5.2 Limit theorem for stochastically stable processes

5.2.1 A uniform weak law of large numbers

The stochastic stability concept will be used for generalizing the theories of chapter 3 and 4 to dynamic models. It will be shown in the next section that if the vector $\underline{x}_j$ of regressors in model (3.1.1) contains lagged dependent variables but the model involved generates a stochastically stable process $(\underline{y}_j)$ with respect to the errors, and if the errors $\underline{u}_j$ are still assumed to be independent, then the results of chapter 3 carries over, as far as convergence in probability and convergence in distribution is concerned. Thus we need a generalisation of both theorem 2.3.4 and the central limit theorem to stochastically stable processes.

It is easily seen that if theorem 2.2.16 carries over for stochastically stable processes, so does theorem 2.3.4. Thus we shall prove first the following analogue of theorem 2.2.16:

Theorem 5.2.1. Let $(\underline{x}_t)$ be a stochastically stable process with respect to a finite dependent[1] base, where the $\underline{x}_t$'s are random vectors in a Euclidean space X. Let the F_t's be the distribution functions of the $\underline{x}_t$'s. Let $\phi(x)$ be a continuous function on X. If

$$\frac{1}{n}\sum_{t=1}^{n} F_t(x) \to G(x) \text{ properly}$$

and

$$\sup_n \frac{1}{n}\sum_{t=1}^{n} E\{\overline{\phi(\underline{x}_t)}^{\circ}\}^{1+\delta} < \infty \quad \text{for some } \delta > 0$$

then

$$\text{plim } \frac{1}{n}\sum_{t=1}^{n} \phi(\underline{x}_t) = \int \phi(x)dG(x)$$

Then we may conclude without further justification the following analogue of theorem 2.3.4:

Theorem 5.2.2. Let $(\underline{x}_t)$ be a stochastically stable process with respect to a finite dependent base, where the $\underline{x}_t$'s are random vectors in a Euclidean space X. Let $F_1(x), F_2(x), \ldots.$ be the distribution functions of the $\underline{x}_j$'s. Let $f(x,\theta)$ be a continuous real function on X x Ⓗ, where Ⓗ is a compact subset of a Euclidean space. If

$$\frac{1}{n}\sum_{t=1}^{n} F_t(x) \to G(x) \text{ properly}$$

and

$$\sup_n \frac{1}{n}\sum_{t=1}^{n} E\{\overline{f(\underline{x}_t,\theta)}^{Ⓗ}\}^{1+\delta} < \infty \text{ for some } \delta > 0$$

then

$$\frac{1}{n}\sum_{t=1}^{n} f(\underline{x}_j,\theta) \to \int f(x,\theta)dG(x) \text{ in pr. uniformly on Ⓗ}$$

Proof of theorem 5.2.1: We need the following lemma's

1) A sequence $(\underline{z}_t)$ of random variables or vectors is called p-dependent if for every t, the sequences $(z_t, \underline{z}_{t-1}, \underline{z}_{t-2}, \ldots..)$ and $(z_{t+p}, z_{t+p+1}, \ldots..)$ are mutually independent. If such a $p<\infty$ exists then $(\underline{z}_t)$ is called finite dependent.

Lemma 5.2.1. Let $\underline{y}_t$ and $\underline{y}_{t,n}$ $(t=1,2,\ldots,n;n=1,2,\ldots)$ be random vectors in a Euclidean space Y such that $G_n(x)=\frac{1}{n}\sum_{t=1}^{n}P(\underline{y}_t\le x)\to G(x)$ properly and $\lim\frac{1}{n}\sum_{t=1}^{n}P(|\underline{y}_t-\underline{y}_{t,n}|>\delta)=0$ for every $\delta>0$. Then

(i) $\frac{1}{n}\sum_{t=1}^{n}P\{\underline{y}_t\le x_1$ and $\underline{y}_{t,n}\le x_2\}\to G(\min(x_1,x_2))$ properly.

(ii) $\frac{1}{n}\sum_{t=1}^{n}P(\underline{y}_{t,n}\le x)\to G(x)$ properly

where $\min(x_1,x_2)$ denotes the vector of minima of corresponding components of x_1 and x_2.

Proof: Let $\underline{y}_1$ and $\underline{y}_2$ be random variables in Y and put $\underline{z}=\underline{y}_2-\underline{y}_1$. For the sake of convenience we assume that $Y=R^1$, but the results below carry over to general Euclidean spaces. We have for every $\delta>0$

$$P\{\underline{y}_1\le x_1 \text{ and } \underline{y}_2\le x_2\}=P\{\underline{y}_1\le\min(x_1,x_2-\underline{z})\}=P\{\underline{y}_1\le\min(x_1,x_2-\underline{z}) \text{ and } |\underline{z}|\leq\delta\}$$
$$+P\{\underline{y}_1\le\min(x_1,x_2-\underline{z}) \text{ and } |\underline{z}|>\delta\}$$

Thus

$$P\{\underline{y}_1\le x_1 \text{ and } \underline{y}_2\le x_2\}\ge P\{\underline{y}_1\le\min(x_1,x_2-\underline{z}) \text{ and } |\underline{z}|\le\delta\}$$
$$\ge P\{\underline{y}_1\le\min(x_1,x_2-\delta) \text{ and } |\underline{z}|\le\delta\}$$
$$=P\{\underline{y}_1\le\min(x_1,x_2-\delta)\}-P\{\underline{y}_1\le\min(x_1,x_2-\delta) \text{ and } |\underline{z}|>\delta\}$$
$$\ge P\{\underline{y}_1\le\min(x_1,x_2-\delta)\}-P\{|\underline{z}|>\delta\}$$

and similarly

$$P\{\underline{y}_1\le x_1 \text{ and } \underline{y}_2\le x_2\}\le P\{\underline{y}_1\le\min(x_1,x_2+\delta)\}+P\{|\underline{z}|>\delta\}$$

Hence,

$$|P\{\underline{y}_1\le x_1 \text{ and } \underline{y}_2\le x_2\}-P\{\underline{y}_1\le\min(x_1,x_2)\}|$$
$$\le P\{\underline{y}_1\le\min(x_1,x_2)+\delta\}-P\{\underline{y}_1\le\min(x_1,x_2)-\delta\}+P\{|\underline{z}|>\delta\}$$

and consequently

$$\limsup_{n\to\infty}\left|\frac{1}{n}\sum_{t=1}^{n}P\{\underline{y}_t\le x_1 \text{ and } \underline{y}_{t,n}\le x_2\}-\frac{1}{n}\sum_{t=1}^{n}P\{\underline{y}_t\le\min(x_1,x_2)\}\right|$$
$$\le\limsup_{n\to\infty}\left|\frac{1}{n}\sum_{t=1}^{n}P\{\underline{y}_t\le\min(x_1,x_2)+\delta\}-\frac{1}{n}\sum_{t=1}^{n}P\{\underline{y}_t\le\min(x_1,x_2)-\delta\}\right|$$
$$+\limsup_{n\to\infty}\frac{1}{n}\sum_{t=1}^{n}P\{|\underline{y}_t-\underline{y}_{t,n}|>\delta\}\le G(+(\min(x_1,x_2)+\delta))-G(-(\min(x_1,x_2)-\delta)),$$

where $G(+x)=\lim_{\varepsilon\downarrow 0}G(x+\varepsilon)$ and $G(-x)=\lim_{\varepsilon\downarrow 0}G(x-\varepsilon)$.

If $\min(x_1,x_2)$ is a continuity point of G then by letting $\delta\downarrow 0$,

$$\left|\frac{1}{n}\sum_{j=1}^{n}P(y_t \leq x_1 \text{ and } y_t \leq x_2) - G_n(\min(x_1,x_2))\right| \to 0 \quad \text{if } n\to\infty,$$

which proves part (i). Part (ii) of the lemma follows from the proof of part(i) by taking $x_1 = \infty$. □

The next lemma is an extension of the weak law of large numbers.

Lemma 5.2.2. Let $\underline{x}_{1,n},\ldots,\underline{x}_{n,n}$, $n=1,2,\ldots$, be for fixed n a double array of m_n-dependent random variables, where (m_n) is a sequence of positive integers such that $m_n=o(n)$. Suppose $E\underline{x}_{t,n}=0$ for $t=1,2,\ldots,n$ and $n=1,2,\ldots$.
If $\frac{1}{n}\sum_{t=1}^{n}E\underline{x}_{t,n}^2$ remains bounded for all $n\geq 1$, then $\operatorname{plim}\frac{1}{n}\sum_{t=1}^{n}\underline{x}_{t,n}=0$

Proof: Since $\underline{x}_{t,n}$ and $\underline{x}_{t+m_n+1,n}$ are independent for $t=1,2,\ldots,n-m_n-1$ it follows that

$$E\left(\frac{1}{n}\sum_{t=1}^{n}\underline{x}_{t,n}\right)^2=\frac{1}{n^2}\sum_{t=1}^{n}E\underline{x}_{t,n}^2+2\frac{1}{n^2}\sum_{t=1}^{n-1}\sum_{j=t+1}^{n}E\underline{x}_{t,n}\underline{x}_{j,n}$$

$$=\frac{1}{n^2}\sum_{t=1}^{n}E\underline{x}_{t,n}^2+2\frac{1}{n^2}\sum_{t=1}^{n-1}\sum_{j=t+1}^{\min(n,t+m_n)}E\underline{x}_{t,n}\underline{x}_{j,n}$$

$$\leq O(n^{-1})+\frac{1}{n^2}\sum_{t=1}^{n-1}\sum_{j=t+1}^{\min(n,t+m_n)}(E\underline{x}_{t,n}^2+E\underline{x}_{j,n}^2)$$

$$=O(n^{-1})+m_n.O(n^{-1})=O(n^{-1})+o(n)O(n^{-1})\to 0 \quad \text{as } n\to\infty \tag{5.2.1}$$

Hence the lemma follows from Chebishev's inequality. □

Lemma 5.2.3. Let $(\underline{y}_t)$ be a stochastically stable process with respect to a finite dependent base, where the $\underline{y}_t$'s are random vectors in a Euclidean space Y. Let the G_t's be the distribution functions of the $\underline{y}_t$'s, respectively. Let $\phi(y)$ be a uniformly bounded continuous function on Y. If

$\frac{1}{n}\sum_{t=1}^{n}G_t(y)\to G(y)$ properly

then

$\operatorname{plim}\frac{1}{n}\sum_{t=1}^{n}\phi(\underline{y}_t)=\int\phi(y)dG(y)$

Proof: Let $(\underline{y}_{t,n})$ be the m_n-approximation of $(\underline{y}_t)$. Thus $\underline{y}_{t,n}$ has the structure

$$\underline{y}_{t,n} = \begin{cases} \phi_{t,n}(\underline{u}_t, \underline{u}_{t-1}, \ldots, \underline{u}_{t-m_n+1}) & \text{if } t \geq m_n \\ y_t & \text{if } 1 \leq t < m_n \end{cases}$$

where $(\underline{u}_t)$ is the finite dependent base involved. But the array $(\underline{y}_{t,n})$ is not necessarily $O(m_n)$-dependent. However the sequence $(\underline{y}^*_{t,n})$ defined by

$$\underline{y}^*_{t,n} = \begin{cases} \phi_{t,n}(\underline{u}_t, \underline{u}_{t-1}, \ldots, \underline{u}_{t-m_n+1}) & \text{if } t \geq m_n \\ 0 & \text{if } t < m_n \end{cases} \tag{5.2.2}$$

is clearly $O(m_n)$-dependent, while

$$\lim_{n\to\infty} \frac{1}{n}\sum_{t=1}^{n} P(|\underline{y}_t - \underline{y}^*_{t,n}| > \delta) \leq \lim \frac{1}{n}\sum_{t=1}^{n} P(|\underline{y}_t - \underline{y}_{t,n}| > \tfrac{1}{2}\delta)$$

$$+ \lim \frac{1}{n}\sum_{t=1}^{n} P(|\underline{y}^*_{t,n} - y_{t,n}| > \tfrac{1}{2}\delta)$$

$$= \lim \frac{1}{n}\sum_{t=1}^{m_n-1} P(|\underline{y}_t| > \tfrac{1}{2}\delta) \leq \lim \frac{m_n}{n} = 0 . \tag{5.2.3}$$

Thus from lemma 5.2.1. and theorem 2.2.10 it follows that under the conditions of the lemma

$$\lim \frac{1}{n}\sum_{t=1}^{n} E|\phi(\underline{y}_t) - \phi(\underline{y}^*_{t,n})| = \int |\phi(y_1) - \phi(y_2)| dG(\min(y_1, y_2)) = 0 \tag{5.2.4}$$

for any uniformly bounded continuous function ϕ on Y. Hence by Chebishev's inequality we conclude:

$$\text{plim}\{\frac{1}{n}\sum_{t=1}^{n}\phi(\underline{y}_t) - \frac{1}{n}\sum_{t=1}^{n}\phi(\underline{y}^*_{t,n})\} = 0 . \tag{5.2.5}$$

Moreover, theorem 2.2.10 and part (ii) of lemma 5.2.1 imply

$$\lim \frac{1}{n}\sum_{t=1}^{n} E\phi(\underline{y}^*_{t,n}) = \int \phi(y) dG(y) , \tag{5.2.6}$$

while obviously from lemma 5.2.2 it follows

$$\text{plim} \frac{1}{n}\sum_{t=1}^{n}\{\phi(\underline{y}^*_{t,n}) - E\phi(\underline{y}^*_{t,n})\} = 0 . \tag{5.2.7}$$

Combining (5.2.5),(5.2.6) and (5.2.7), the lemma follows. □

This lemma shows that (2.2.13) in the proof of theorem 2.2.16 holds for stochastically stable processes with respect to a finite dependent base, hence theorem 2.2.16 itself carries over. Thus theorem 5.2.1 is proved now and so is theorem 5.2.2. □

5.2.2 Martingales

In the sections 3.1, 3.2 and 3.3 we have applied the central limit theorem to sums

$$\underline{s}_n=\sum_{t=1}^{n}\underline{v}_j$$

with

$$\underline{v}_j=\begin{cases} \underline{u}_j(\partial/\partial\theta)g(\underline{x}_j,\theta_0)'\zeta & \text{in the case of section 3.1,} \\ \frac{1}{\gamma^2}\rho'\left(\frac{\underline{u}_j}{\gamma}\right)(\partial/\partial\theta)g(\underline{x}_j,\theta_0)'\zeta & \text{in the case of section 3.2 ,} \\ \frac{1}{\gamma^2}\rho'\left(\frac{\underline{u}_j}{\gamma s(\underline{x}_j)}\right)\frac{(\partial/\partial\theta)g(\underline{x}_j,\theta_0)'\zeta}{s(\underline{x}_j)} & \text{in the case of section 3.3,} \end{cases}$$

where ζ was some non-random vector in R^q. Since by assumption 3.1.1 these $\underline{v}_j$'s were independent and by the assumptions 3.1.2 and 3.2.1, respectively, $E\underline{v}_j=0$ it follows that for each n,

$$E_{n-1}\underline{s}_n=E\{\underline{s}_n|\underline{v}_1,\ldots,\underline{v}_{n-1}\}=\sum_{j=1}^{n-1}\underline{v}_j+E\underline{v}_n=\underline{s}_{n-1} \quad \text{a.s.,} \tag{5.2.8}$$

or in words the conditional expectation of $\underline{s}_n$, given $\underline{v}_1,\ldots,\underline{v}_{n-1}$, equals $\underline{s}_{n-1}$ almost surely. This makes the sequence $(\underline{s}_n)$ a Martingale [see Chung (1974, chapter 9) for a precise definition of martingale] and the sequence $(\underline{v}_j)$ a sequence of martingale differences:

$$E_{n-1}\underline{v}_n=E(\underline{v}_n|\underline{v}_1,\ldots,\underline{v}_{n-1})=0 \quad \text{a.s. .} \tag{5.2.9}$$

Now suppose that the vector $\underline{x}_j$ in model (3.1.1) contains lagged dependent variables:

$$\underline{x}_j'=(\underline{y}_{j-1},\underline{y}_{j-2},\ldots,\underline{y}_{j-p_1},\underline{x}^*_{1,j},\ldots,\underline{x}^*_{p_2,j}),p_1+p_2=p \ , \tag{5.2.10}$$

say, and put

$$\underline{x}_j^{*\prime}=(\underline{x}^*_{1,j},\ldots,\underline{x}^*_{p_2,j}) \ . \tag{5.2.11}$$

Moreover, assume:

Assumption 5.2.1. The sequence of pairs $(\underline{u}_j,\underline{x}_j^*)$ is independent. For each j, $\underline{u}_j$ and $\underline{x}_j^*$ are mutually independent and the sequence $(\underline{u}_j,\underline{x}_j^*)$, $j\geq 1$, is independent of $\underline{y}_0,\underline{y}_{-1},\ldots,\underline{y}_{-p_1+1}$.

Then obviously (5.2.8) and (5.2.9) are still satisfied. Moreover, if the sequence $(\underline{y}_j)$ generated by model (3.1.1) with (5.2.10) is stochastically stable with respect to the base $(\underline{u}_j,\underline{x}_j^*)$, then under the assumptions 3.1.1 and 5.2.1 the sequence of pairs $(\underline{u}_j,\underline{x}_j)$ is also stochastically stable with respect to the base

$(\underline{u}_j,\underline{x}_j^*)$ because, putting

$$\underline{x}_{j,n}' = (\underline{y}_{j-1,n},\ldots,\underline{y}_{j-p_1,n},\underline{x}_{1,j}^*,\ldots,\underline{x}_{p_2,j}^*)\ , \tag{5.2.12}$$

where $(\underline{y}_{j,n})$ is the m_n-approximation of $(\underline{y}_j)$, we have for any $\delta>0$, using inequality (5.1.32),

$$\frac{1}{n}\sum_{j=1}^n P\{|(\underline{u}_j,\underline{x}_j)-(\underline{u}_j,\underline{x}_{j,n})|>\delta\} = \frac{1}{n}\sum_{j=1}^n P(\sum_{i=1}^{p_1}(\underline{y}_{j-i}-\underline{y}_{j-i,n})^2>\delta^2)$$

$$\leq \sum_{i=1}^{p_1}\frac{1}{n}\sum_{j=1}^n P\{|\underline{y}_{j-i}-\underline{y}_{j-i,n}|>\frac{\delta}{\sqrt{p_1}}\}\rightarrow 0\ . \tag{5.2.13}$$

Thus for generalizing the results of chapter 3 to dynamic models we need a central limit theorem for a sequence of stochastically stable martingale differences. But such a central limit theorem can easily be derived from the central limit theorem for martingales of Brown (1971):

Theorem 5.2.3. Let $(\underline{s}_n)$ be a martingale sequence and let $\underline{v}_n=\underline{s}_n-\underline{s}_{n-1};\underline{s}_0=0$. Put

$$\underline{w}_n^2=\sum_{t=1}^n E_{t-1}\underline{v}_t^2\ ;\ \sigma_n^2=E\underline{w}_n^2=\sum_{t=1}^n E\underline{v}_t^2\ . \tag{5.2.14}$$

If

$$\frac{\underline{w}_n^2}{\sigma_n^2}\rightarrow 1 \quad \text{in pr.} \tag{5.2.15}$$

and

$$\sigma_n^{-2}\sum_{t=1}^n E\underline{v}_t^2 I(|\underline{v}_t|\geq\varepsilon\sigma_n)\rightarrow 0^{*)} \tag{5.2.16}$$

for all $\varepsilon>0$, then

$$\frac{\underline{s}_n}{\sigma_n}\rightarrow N(0,1) \text{ in distr.}.$$

Proof: see Brown (1971)

---

*) I(A) denotes the indicator function of the set A: I(A)=1 if a∈A and I(A)=0 if a∉A.

5.2.3 Central limit theorem for stochastically stable martingale differences

Let $(\underline{y}_t)$ be the stochastically stable process as defined by definition 5.1.1 and suppose that the base $(\underline{u}_t)$ involved is i.i.d. with common distribution function H. Moreover, suppose that $(\underline{u}_t)_{t=1}^{\infty}$ is independent of the initial values $\underline{y}_0, \underline{y}_{-1}, \dots, \underline{y}_{-k+1}$. Put

$$\underline{v}_t = \phi_1(\underline{u}_t)\,\phi_2(\underline{y}_{t-1}) \tag{5.2.17}$$

where $\phi_1(u)$ is a continuous function on U such that $E\phi_1(\underline{u}_t)=0$ and $0<E\phi_1(\underline{u}_t)^2<\infty$ and ϕ_2 is a continuous function on Y. Then obviously

$$E_{t-1}\underline{v}_t=\phi_2(\underline{y}_{t-1})E\phi_1(\underline{u}_t)=0 \quad \text{a.s.}, \tag{5.2.18}$$

hence the $\underline{v}_t$'s are martingale differences.

We shall now set forth further conditions such that this sequence $(\underline{v}_t)$ fits the conditions of Brown's martingale central limit theorem, so that $\frac{1}{\sqrt{n}}\sum_{t=1}^{n}\underline{v}_t$ converges in distribution to the normal (with zero mean).

Since the base $(\underline{u}_t)$ is independent, it follows that

$$E_{t-1}\underline{v}_t^2=\phi_2(\underline{y}_{t-1})^2E\,\phi_1(\underline{u}_t)^2=\phi_2(\underline{y}_{t-1})^2E\,\phi_1(\underline{u}_1)^2 . \tag{5.2.19}$$

Thus

$$\begin{aligned}\underline{w}_n^2&=\sum_{t=1}^{n}E_{t-1}\underline{v}_t^2=\{E\phi_1(\underline{u}_1)^2\}\sum_{t=1}^{n}\phi_2(\underline{y}_{t-1})^2 ,\\ \sigma_n^2&=\sum_{t=1}^{n}E\underline{v}_t^2=\{E\phi_1(\underline{u}_1)^2\}\sum_{t=1}^{n}E\phi_2(\underline{y}_{t-1})^2 .\end{aligned} \tag{5.2.20}$$

Let $(F_t(y))$ be the sequence of distribution functions of the $\underline{y}_t$'s and suppose

$$\frac{1}{n}\sum_{t=1}^{n}F_t(y)\rightarrow G(y) \text{ properly} \tag{5.2.21}$$

and

$$\sup_n \frac{1}{n}\sum_{t=1}^{n}E\{|\phi_2(\underline{y}_t)|\}^{2+\delta}<\infty \quad \text{for some } \delta>0 . \tag{5.2.22}$$

Then by theorem 5.2.1 and theorem 2.2.15

$$\operatorname{plim}\frac{1}{n}\underline{w}_n^2=\int\phi_2(y)^2dG(y)E\phi_1(\underline{u}_1)^2=\lim\frac{1}{n}\sigma_n^2 , \tag{5.2.23}$$

hence (5.2.15) is satisfied, provided that $\int\phi_2(y)^2dG(y)>0$. Moreover, in the case (5.2.17) we have for any fixed $a>0$,

$$\limsup_{n\to\infty} \frac{1}{n}\sum_{t=1}^{n} E\underline{v}_t^2\, I(|\underline{v}_t|>\varepsilon\sigma_n) \le \limsup_{n\to\infty} \frac{1}{n}\sum_{t=1}^{n} E\underline{v}_t^2\, I(|\underline{v}_t|>a)$$

$$= \lim_{n\to\infty} \int_{\{|\phi_1(u)\phi_2(y)|>a\}} \phi_1(u)^2\phi_2(y)^2 d\left(\frac{1}{n}\sum_{t=1}^{n} F_{t-1}(y)\right)dH(u) =$$

$$= \lim_{n\to\infty} \int_{\{|\phi_1(u)\phi_2(y)|>a \text{ and } |\phi_1(u)|\le\sqrt{a}\}} \phi_1(u)^2\phi_2(y)^2 d\left(\frac{1}{n}\sum_{t=1}^{n} F_{t-1}(y)\right)dH(u) +$$

$$+ \lim_{n\to\infty} \int_{\{|\phi_1(u)\phi_2(y)|>a \text{ and } |\phi_1(u)|>\sqrt{a}\}} \phi_1(u)^2\phi_2(u)^2 d\left(\frac{1}{n}\sum_{t=1}^{n} F_{t-1}(y)\right)dH(u) \le$$

$$\le \lim_{n\to\infty} \int_{\{|\phi_2(y)|>\sqrt{a}\}} \phi_2^2(y) d\left(\frac{1}{n}\sum_{t=1}^{n} F_{t-1}(y)\right) \int \phi_1(u)^2 dH(u) +$$

$$+ \lim_{n\to\infty} \int \phi_2^2(y) d\frac{1}{n}\sum_{t=1}^{n} F_{t-1}(y) \int_{\{|\phi_1(u)|>\sqrt{a}\}} \phi_1(u)^2 dH(u) =$$

$$= \int_{\{|\phi_2(y)|>\sqrt{a}\}} \phi_2(y)^2 dG(y) \int \phi_1(u)^2 dH(u)$$

$$+ \int \phi_2(y)^2 dG(y) \int_{\{|\phi_1(u)|>\sqrt{a}\}} \phi_1(u)^2 dH(u). \qquad (5.2.24)$$

The last equality in (5.2.24) follows from theorem 2.2.15 and the following lemma.

Lemma 5.2.4. Let the conditions of theorem 2.2.15 be satisfied. Then

$$\int_{\{\phi(x)\le a\}} \phi(x)dF_n(x) \to \int_{\{\phi(x)\le a\}} \phi(x)dF(x)$$

for any continuity point a of the function $\int_{\{\phi(x)\le a\}} dF(x)$

Proof: Put $G_n(a)=\int_{\{\phi(x)\le a\}} dF_n(x)$, $G(a)=\int_{\{\phi(x)\le a\}} dF(x)$. Then for any uniformly bounded continuous function ψ we have

$$\int \psi(a)dG_n(a) = \int \psi(\phi(x))dF_n(x) \to \int \psi(\phi(x))dF(x) = \int \psi(a)dG(a) ,$$

hence by theorem 2.2.10,

$$G_n(a) \to G(a) \text{ properly}. \qquad (5.2.25)$$

Next put

$$\phi_a(x)=\begin{cases}\phi(x) & \text{if } \phi(x)\le a\\ a & \text{if } \phi(x)>a.\end{cases} \tag{5.2.26}$$

Then by theorem 2.2.15 and (5.2.25)

$$\int_{\{\phi(x)\le a\}}\phi(x)dF_n(x)=\int\phi_a(x)dF_n(x)-a\int_{\{\phi(x)>a\}}dF_n(x)=\int\phi_a(x)dF_n(x)-a(1-G_n(a))$$

$$\to\int\phi_a(x)dF(x)-a(1-G(a))=\int_{\{\phi(x)\le a\}}\phi(x)dF(x) \tag{5.2.27}$$

which proves the lemma. □

<u>Remark</u>. Obviously, under the conditions of the lemma we also have:

$$\int_{\{\phi(x)>a\}}\phi(x)dF_n(x)\to\int_{\{\phi(x)>a\}}\phi(x)dF(x)$$

Since all the integrals in (5.2.24) converge, we can make (5.2.24) arbitrarily small by increasing a. Hence (5.2.16) holds for the case (5.2.17). So we have proved by now:

<u>Theorem 5.2.4</u>. Let $(\underline{y}_t)$ be the stochastically stable process as defined in definition 5.1.1, let the base $(\underline{u}_t)$ involved be i.i.d. and let $(\underline{u}_t)_{t=1}^{\infty}$ be independent of $\underline{y}_0,\underline{y}_{-1},\ldots,\underline{y}_{-k+1}$. Let $(F_t(y))$ be the sequence of distribution functions of the $\underline{y}_t$'s and let $H(u)$ be the distribution function of the $\underline{u}_t$'s. Define $\underline{v}_t=\phi_1(\underline{u}_t)\phi_2(\underline{y}_{t-1})$, where ϕ_1 is a continuous function on U such that
$E\phi_1(\underline{u}_t)=0$, $0<E\phi(\underline{u}_t)^2<\infty$
and ϕ_2 is a continuous function on Y such that

$$\sup_n \frac{1}{n}\sum_{t=1}^{n}E\{\overline{\phi_2(\underline{y}_t)}^{\,\circ}\}^{2+\delta}<\infty$$

for some $\delta>0$.
If $\frac{1}{n}\sum_{t=1}^{n}F_t(y)\to G(y)$ properly, and $\int\phi_2(y)^2dG(y)>0$, then
$\frac{1}{\sqrt{n}}\sum_{t=1}^{n}\underline{v}_t\to N[0,\int\phi_1(u)^2dH(u)\int\phi_2(y)^2dG(y)]$ in distr.

This theorem is especially appropriate for generalizing the theory of our sections 3.1, 3.2 and 4.1 to dynamic models, but not for the weighted M-estimation method and the minimum information estimation method. For the latter cases we need the following central limit theorem.

Theorem 5.2.5. Let the sequences $(\underline{y}_t)$ and $(\underline{u}_t)$ be as in theorem 5.2.4. Let now $\underline{v}_t=\phi(\underline{u}_t,\underline{y}_{t-1})$, where ϕ is a continuous function on UxY such that

$E_{t-1}\underline{v}_t=\int\phi(u,\underline{y}_{t-1})dH(u)=0$ a.s., where H is the distribution function of the $\underline{u}_t$'s. Suppose there is a continuous function $\psi(y)$ on Y such that

$|\phi(u,y)|\leq\psi(y)$ for all u in U and y in Y

and

$$\sup_n \frac{1}{n}\sum_{t=1}^n E\psi(\underline{y}_{t-1})^{2+\delta}<\infty \text{ for some } \delta>0.$$

If $\frac{1}{n}\sum_{t=1}^n F_t(y)\to G(y)$ properly, where F_t is the distribution function of $\underline{y}_t$, then

$$\frac{1}{\sqrt{n}}\sum_{t=1}^n \underline{v}_t\to N(0,\iint\phi(u,y)^2dH(u)dG(y)) \text{ in distr.}$$

provided that $\iint\phi(u,y)^2dH(u)dG(y)>0$.

Proof: Obviously we have

$$\frac{1}{n}\sigma_n^2=\frac{1}{n}\sum_{t=1}^n E\underline{v}_t^2=\frac{1}{n}\sum_{t=1}^n\iint\phi(u,y)^2dH(u)dF_{t-1}(y)\to\sigma^2:=\iint\phi(u,y)^2dH(u)dG(y)>0,$$

while for any constant K

$$\left|\frac{1}{n}\sum_{t=1}^n E_{t-1}\underline{v}_t^2-\sigma^2\right|\leq\int\left|\frac{1}{n}\sum_{t=1}^n\phi(u,\underline{y}_{t-1})^2-\int\phi(u,y)^2dG(y)\right|dH(u)$$

$$\leq\sup_{\{|u|\leq K\}}\left|\frac{1}{n}\sum_{t=1}^n\phi(u,\underline{y}_{t-1})^2-\int\phi(u,y)^2dG(y)\right|$$

$$+\frac{1}{n}\sum_{t=1}^n\psi(\underline{y}_{t-1})^2\int_{\{|u|>K\}}dH(u)+\int\psi(y)^2dG(y)\int_{\{|u|>K\}}dH(u)$$

$$\to 2\int\psi(y)^2dG(y)\int_{\{|u|>K\}}dH(u) \text{ in pr.,} \tag{5.2.28}$$

where we have applied the theorems 5.2.1 and 5.2.2 (the latter with $\textcircled{H}=\{|u|\leq K\}$). Since the right hand side of (5.2.28) can be made arbitrarily small by increasing K, it follows that

$$\text{plim}\frac{1}{n}\sum_{t=1}^n E_{t-1}\underline{v}_t^2=\sigma^2>0. \tag{5.2.29}$$

Thus condition (5.2.15) is satisfied. Next we observe that

$$\frac{1}{n}\sum_{t=1}^n E\underline{v}_t^2\, I(|\underline{v}_t|\geq\varepsilon\sigma_n)=\frac{1}{n}\sum_{t=1}^n\iint_{\{|\phi(u,y)|\geq\varepsilon\sigma_n\}}\phi(u,y)^2dH(u)dF_{t-1}(y)\leq\frac{1}{n}\sum_{t=1}^n\int_{\{|\psi(y)|\geq\varepsilon\sigma_n\}}\psi(y)^2dF_{t-1}(y),$$

so that from lemma 5.2.4 and the fact that $\sigma_n \to \infty$ it follows that also condition (5.2.16) is satisfied. This completes the proof.

5.3 Dynamic nonlinear regression models and implicit structural equations

5.3.1 Dynamic nonlinear regression models

In this section we shall generalize the convergence in probability results of the chapters 3 and 4 to models with lagged dependent variables but i.i.d. errors $\underline{u}_j$. In the case of chapter 3 we still adopt model (3.1.1) but now the regressors are of the type (5.2.10). Of course in this case assumption 3.1.1 has to be modified because the regressors $\underline{x}_j$ can no longer be independent. Thus we assume

Assumption 5.3.1. Assumption 3.1.1 is satisfied except that now the regressors $\underline{x}_j$ are no longer independent.

However, obviously under the assumptions 5.2.1 and 5.3.1 we still have that $\underline{x}_j$ and $\underline{u}_j$ are mutually independent. Now suppose in addition that

Assumption 5.3.2. The sequence $(\underline{y}_j)$ generated by model (3.1.1) [with regressors of the type (5.2.9)] is stochastically stable with respect to the base $((\underline{u}_j,\underline{x}_j^*))$.

From (5.2.12) we have already seen that

Lemma 5.3.1. Under the assumptions 5.2.1, 5.3.1 and 5.3.2, the sequence $(\underline{x}_j)$ of regressors and the sequence of pairs $(\underline{u}_j,\underline{x}_j)$ are both stochastically stable with respect to the base $((\underline{u}_j,x_j^*))$.

All the convergence in probability results in chapter 3 were based on theorem 2.3.4. But in view of theorem 5.2.2 and lemma 5.3.1, refering to theorem 5.2.2 instead of theorem 2.3.4 would yield the same conclusions. Moreover, where we refered to the central limit theorem 2.4.5 we could refer to the central limit theorems 5.2.4 or 5.2.5 without making additional assumptions. Thus we have:

Theorem 5.3.1. If assumption 3.1.1 is replaced with the assumptions 5.2.1, 5.3.1 and 5.3.2 then:

1) the theorems 3.1.3 through 3.1.6 carry over.
2) theorem 3.2.8 carries over.
3) the theorems 3.3.1 through 3.3.4 carry over, except that now the conclusions "a.s." have to be changed in: "in pr."
4) the same applies for the theorems 3.4.1 and 3.4.2, while also theorem 3.4.3 carries over.

However, with respect to the parts 2) of this theorem it must be noted that the assumptions 3.1.11 and 3.2.2 may fail to hold if some moments of the errors are infinite. For example consider the simple linear autoregression

$$\underline{y}_j = \theta_0 \underline{y}_{j-1} + \underline{u}_j \ , \ j=0,\pm 1,\pm 2,\ldots \ , \ 0<\theta_0<1$$

where the $\underline{u}_j$'s are random drawings from the Cauchy (0,1) distribution. Then $\underline{y}_j = \sum_{k=0}^{\infty} \theta_0^k \underline{u}_{j-k}$, hence

$$E e^{it\underline{y}_j} = \prod_{k=0}^{\infty} E e^{it\theta_0^k \underline{u}_{j-k}} = \prod_{k=0}^{\infty} e^{-\theta_0^k |t|} = e^{-\frac{1}{1-\theta_0}|t|} ,$$

so that $\underline{y}_j$ is also Cauchy distributed. In this case $\underline{x}_j = \underline{y}_{j-1}$, $g(\underline{x}_j,\theta) = \theta \underline{y}_{j-1}$ and $(\partial/\partial\theta) g(\underline{x}_j,\theta) = \underline{y}_{j-1}$, so that assumption 3.1.11 cannot be satisfied.

5.3.2 Dynamic nonlinear implicit structural equations

Generalisation of the results of chapter 4 to the dynamic case is also easy, provided that we still assume that the errors $\underline{u}_j$ are i.i.d. Thus we modify the assumptions 4.1.1 and 4.1.2 as follows:

> Assumption 5.3.3. Assumption 4.1.1 is satisfied, except that now the sequence $(\underline{z}_j)$ is stochastically stable with respect to an independent base $(\underline{u}_j,\underline{v}_j)$, say, where $(\underline{u}_j,\underline{v}_j)_{j=1}^{\infty}$ is independent of the initial $\underline{z}_j$'s, $j \leq 0$.

> Assumption 5.3.4. Assumption 4.1.2 is satisfied, except that now the sequence of pairs $(\underline{w}_j,\underline{z}_j)$ is stochastically stable with respect to an independent base $(\underline{u}_j,\underline{v}_j)$, say, where $(\underline{u}_j,\underline{v}_j)_{j=1}^{\infty}$ is independent of the initial $\underline{z}_j$'s, $j \leq 0$.

Then it follows from theorem 5.2.4 that (4.1.24) carries over, while all the convergence in probability results in theorem 4.1.3 now follow from theorem 5.2.2 instead of theorem 2.3.4. Thus we have:

> Theorem 5.3.2. If the assumptions 4.1.1 and 4.1.2 are replaced with the assumptions 5.3.3 and 5.3.4, then theorem 4.1.3 carries over.

Notice that the assumptions 5.3.3 and 5.3.4 allow for lagged dependent variables to be used as instrumental variables, so that we now have answered the question put by Gallant (1977) who stated:"... there are no results available in the probability literature, to the author's knowledge, which yields the conclusions (...) for lagged endogenous instrumental variables generated by a system of simultaneous, nonlinear, implicit equations".

Furthermore, using theorem 5.2.2 instead of theorem 2.3.4 we see that

Theorem 5.3.3. If assumption 4.1.1 is replaced with assumption 5.3.3 then the conclusions of the theorems 4.2.1 and 4.3.1 now become $\underline{\theta}_n^0 \rightarrow \theta_0$ in pr. and $\underline{\theta}_n^0(\gamma)\rightarrow\theta_0$ in pseudo-uniformly on Γ, respectively.

Under the assumptions 5.3.3 and 4.2.1 we still have that the random variables $\sin(t\underline{u}_j w(\underline{z}_j))$ have zero expectations and covariances for each t. Thus (4.2.21) carries over, so that also now the random variables $\underline{v}_n = \int_0^\infty n\, \underline{s}_n(t,\theta_0)^2\phi(t)dt$ are stochastically bounded. Applying theorem 5.3.2 and lemma 4.2.1 we then conclude that the righthand side of (4.2.18) converges in pr. to zero and hence that (4.2.17) carries over. Furthermore, since the random variables $\int_0^\infty \sin(t\underline{u}_j h(\underline{z}_j))c_i(t,\theta_0)\frac{1}{\gamma}\phi(t/\gamma)dt$ satisfy the conditions of theorem 5.2.5 it follows that both (4.2.11) and (4.3.49) carry over. Thus without further discussion we may now conclude that:

Theorem 5.3.4. If assumption 4.1.1 is replaced with assumption 5.3.3 then the theorems 4.2.2, 4.3.2 and 4.3.5 carry over.

Finally, since the $\underline{z}_j$'s in lemma 4.4.1 may be any random vectors, it follows that

Theorem 5.3.5. If assumption 4.1.1 is replaced with assumption 5.3.3 then the conclusion of theorem 4.4.3 becomes $\tilde{\underline{\theta}}_n^*(\gamma)\rightarrow\theta_0$ in pr. pseudo-uniformly on Γ.

5.4 Remarks on the stochastic stability concept

Definition 5.1.1 considers only processes $(\underline{y}_t)$ for $t \geq 1$ (or more generally for t bounded from below). But if t can also be 0,-1,-2,..... then by repeated substitution, (5.1.12) becomes

$$\underline{y}_t = \phi_t^*(\underline{u}_t, \underline{u}_{t-1}, \underline{u}_{t-2}, \ldots\ldots).$$

In that case it would be natural to redefine (5.1.14) as follows

$$\underline{y}_{t,n} = \phi_{t,n}^*(\underline{u}_t, \underline{u}_{t-1}, \ldots, \underline{u}_{t-m_n+1}).$$

In order that this case is also covered by the stochastic stability concept, we therefore define:

> **Definition 5.4.1.** Let $(\underline{y}_t)$, $t=0,\pm1,\pm2,\ldots\ldots$ be a sequence of random vectors in a Euclidean space Y. Suppose that
>
> $$\underline{y}_t = \phi_t(\underline{u}_t,\underline{u}_{t-1},\underline{u}_{t-2},\ldots\ldots),$$
>
> where the $\underline{u}_t$'s are random vectors in a Euclidean space U and the ϕ_t's are Borel measurable mappings from $\overset{\infty}{\underset{i=1}{X}}$ U into Y. Let (m_n) be a sequence of positive integers such that $m_n=o(n)$. Consider a double array $\underline{y}_{t,n}$, $n=1,2\ldots$, $t=0,\pm1,\pm2,\ldots$ with the structure
>
> $$\underline{y}_{t,n}=\phi_{t,n}(\underline{u}_{t-1},\ldots\ldots\underline{u}_{t-m_n}+1),$$
>
> where the $\phi_{t,n}$'s are Borel measurable mappings from $\overset{m_n}{\underset{i=1}{X}}$ U into Y. If for such a double array, (5.1.15) is satisfied for any $\delta>0$, then $(\underline{y}_t)$ is also said to be stochastically stable with respect to the base $(\underline{u}_t)$.

Obviously, for these stochastically stable processes all the limit theorems of section 5.2 carry over.

As an example of such a process, consider

$$\underline{y}_t = \sum_{j=0}^{\infty} \alpha_j\underline{u}_{t-j} ,$$

where the $\underline{u}_t$'s are identically distributed random variables (or vectors) and the scalars α_j's are such that the process $(\underline{y}_t)$ is stationary. Put

$$\underline{y}_t^{(m)} = \sum_{j=0}^{m-1} \alpha_j\underline{u}_{t-j} .$$

Then obviously

$$\underline{y}_t - \underline{y}_t^{(m)} = \sum_{j=m}^{\infty} \alpha_j\underline{u}_{t-j} \to 0 \text{ a.s. as } m \to \infty ;$$

hence also $\underset{m\to\infty}{\text{plim}}(\underline{y}_t-\underline{y}_t^{(m)})=0$ or with other words,

$$\lim_{m\to\infty} P(|\underline{y}_t-\underline{y}_t^{(m)}| > \delta)=0 \text{ for any } \delta > 0 .$$

Because of stationarity, $P(|\underline{y}_t-\underline{y}_t^{(m)}|> \delta)$ is independent of t, hence

$$\frac{1}{n}\sum_{j=1}^{n}P(|\underline{y}_t-\underline{y}_t^{(m_n)}|> \delta) \to 0 \text{ as } n\to\infty \text{ and } m_n \to \infty ,$$

which was to be shown.

Stochastic stability is only a meaningful concept if the base involved has particular properties. For, any sequence of random vectors or variables is stochastically stable with respect to itself. In section 5.2 we have proved limit theorems for stochastically stable processes with respect to a finite dependent or independent base. But finite dependence (hence also independence) is a special case of ϕ-mixing [see for example Billingsley (1969) for a definition and some properties of ϕ-mixing processes]. It might be possible that some of our limit theorems carry over for stochastically stable processes with respect to more general classes of ϕ-mixing processes.

6 SOME APPLICATIONS

In this chapter we present two applications of robust M-estimation and an application of the minimum information estimation method. The two applications of robust M-estimation involve linear models. The first one originates from the research program of the Foundation for Economic Research of the University of Amsterdam, the second one comes from Christopher A. Sims, who suggested this application and kindly provided the data. The linearity of the models is no serious limitation, because the main reason for these examples is to show that non-normal errors do occur in practice and that in these cases our robust M-estimation method may be more efficient.

The minimum information estimation method is applied to a small structural equation, nonlinear in the parameters, which relates the money stock, relative to net national income, to the money market interest rate. The purpose of this example is merely to show how the estimation method involved can be carried out, using only one instrumental variable.

6.1 Applications of robust M-estimation

6.1.1 Municipal expenditure

In the contract research program of the Foundation for Economic Research of the University of Amsterdam we have built a model*) which explains expenditure of the municipality of Amsterdam during the period 1951-1974. One of the equations involved dealt with the logarithm of the expenditure at constant prices (denoted by $\underline{y}$) of the department of sanitation. We used the following explanatory variables:

$\underline{x}_1 = \frac{\text{total expenditure} - \text{total income}}{\text{total expenditure}}$, lagged three years. [In connection with budget constraints]

$\underline{x}_2$= the logarithm of the total number of houses, lagged three years**).

$\underline{x}_3$= the logarithm of the centre-function-index, lagged three years**). The index involved measures the extent to which the city of Amsterdam is visited by non-inhabitants. It is defined as the annual number of picture-goers of Amsterdam cinemas divided by the number of inhabitants of Amsterdam, relative to the same quantity for the Netherlands.

---

*) See Bierens, Stoffel and Poelert (1980) for more applications of robust M-estimation. The model reported here is only a preliminary result.

**)These lags are estimated in a first stage, using the method described in Bierens (1980).

$\underline{x}_4$= the number of frost-days [In connection with snow-clearing].
The model was specified as:

$$\underline{y}_j=\theta_1\underline{x}_{1,j}+\theta_2\underline{x}_{2,j}+\theta_3\underline{x}_{3,j}+\theta_4\underline{x}_{4,j}+\theta_5+\underline{u}_j,$$

where j is the index of the year and $\underline{u}_j$ is a disturbance term.
The parameter θ_1 was expected to be negative and the parameters θ_2 through θ_4 were expected to be positive. Putting

$$\underline{x}_j'=(\underline{x}_{1,j},\underline{x}_{2,j},\underline{x}_{3,j},\underline{x}_{4,j},1),\theta_0'=(\theta_1,\theta_2,\theta_3,\theta_4,\theta_5),$$

$$\underline{X}=\begin{pmatrix}\underline{x}_1'\\ \vdots\\ \underline{x}_n\end{pmatrix},\ \underline{y}=\begin{pmatrix}\underline{y}_1\\ \vdots\\ \underline{y}_n\end{pmatrix},\ \underline{u}=\begin{pmatrix}\underline{u}_1\\ \vdots\\ \underline{u}_n\end{pmatrix},$$

where n is the number of observations, the model can now be written in matrix notation as

$$\underline{y}=\underline{X}\theta_0+\underline{u}\ .$$

We recall that under the conditions for asymptotic normality of the least squares estimator $\hat{\underline{\theta}}_n$ [see section 3.1] we have

$$\sqrt{n}(\hat{\underline{\theta}}_n-\theta_0)\to N(0,\sigma^2(\text{plim}\ \tfrac{1}{n}\ \underline{X}'\underline{X})^{-1})\ \text{in distr.,} \tag{6.1.1}$$

where σ^2 is the variance of the $\underline{u}_j$'s.

We started estimation by OLS, which yielded the following results:

Table 6.1.1: OLS-results

	θ_1	θ_2	θ_3	θ_4	θ_5
coefficients	-1.624	1.022	0.950	0.004	-8.068
standard errors	0.794	0.254	0.190	0.001	3.122

R^2	0.838	(multiple correlation coefficient)
S.E.	0.064	(residual standard error corrected for loss of degrees of freedom)
D.W.	1.061	(Durbin-Watson statistic)
n	24	(number of observations)
W.R.K.	3.36	(weighted residual kurtosis)

The statistic WRK is defined by (3.4.24), where the function s involved was chosen:

$$s(\underline{x}_j)=1+|\underline{x}_{1,j}|+|\underline{x}_{2,j}|+|\underline{x}_{3,j}|+|\underline{x}_{4,j}|\ .$$

In view of remark 3 at theorem 3.1.6 we concluded that the error distribution probably has a finite variance, so that $\sigma^2 < \infty$, but in view of theorem 3.4.3, since WRK > 0, we also concluded that the above results may be improved by robust M-estimation.

We recall that under the conditions for asymptotic normality of the robust M-estimator $\tilde{\underline{\theta}}_n(\gamma)$ [see section 3.2] we have:

$$\sqrt{n}(\tilde{\underline{\theta}}_n(\gamma) - \theta_0) \to N(0, h(\gamma)(\text{plim } \tfrac{1}{n}\underline{X}'\underline{X})^{-1}) \text{ in distr.}, \qquad (6.1.2)$$

where

$$h(\gamma) = \frac{\gamma^2 \int \rho'(u/\gamma)^2 f(u)du}{\{\int \rho''(u/\gamma)f(u)du\}^2}$$

with γ a scaling parameter, ρ a symmetric unimodal density and f the error density.

Since WRK > 0, we may expect that also plim WRK > 0. In that case we have:

$h(\gamma) \uparrow \sigma^2$ as $\gamma \to \infty$.

Thus $h(\gamma) < \sigma^2$ for "large" γ, and for such γ the robust M-estimator $\tilde{\underline{\theta}}_n(\gamma)$ is therefore asymptotically more efficient than the least squares estimator $\hat{\underline{\theta}}_n$.

We have seen in section 3.2.5 that the random function

$$\underline{h}_n(\gamma, \underline{\theta}_n^*) = \frac{\underline{h}_{1,n}(\gamma, \underline{\theta}_n^*)}{\underline{h}_{2,n}(\gamma, \underline{\theta}_n^*)^2}, \qquad (6.1.3)$$

with $\underline{\theta}_n^*$ a consistent estimator of θ_0 and

$$\underline{h}_{1,n}(\gamma,\theta) = \gamma^2 \tfrac{1}{n}\textstyle\sum_{j=1}^n \rho'\left(\frac{\underline{y}_j - \underline{x}_j'\theta}{\gamma}\right)^2 ,$$

$$\underline{h}_{2,n}(\gamma,\theta) = \tfrac{1}{n}\textstyle\sum_{j=1}^n \rho''\left(\frac{\underline{y}_j - \underline{x}_j'\theta}{\gamma}\right) ,$$

may serve as a uniformly consistent estimator of the function $h(\gamma)$ on a compact subinterval Γ of $(0,\infty)$. In the present case, where $\sigma^2 = \int u^2 f(u)du < \infty$, we may even let $\Gamma = [\gamma_*, \infty)$ with $\gamma_* > 0$. Moreover, we have seen in section 3.2.6 that the optimal scaling parameter γ_0, say [satisfying $h(\gamma_0) = \inf_{\gamma\in\Gamma} h(\gamma)$], can be consistently estimated by a statistic $\tilde{\underline{\gamma}}_n$ satisfying

$$\tilde{\underline{\gamma}}_n \in \Gamma,\ \underline{h}_n(\tilde{\underline{\gamma}}_n, \underline{\theta}_n^*) = \inf_{\gamma\in\Gamma} \underline{h}_n(\gamma, \underline{\theta}_n^*),$$

and that the two-stage M-estimator $\tilde{\underline{\theta}}_n(\tilde{\underline{\gamma}}_n)$ has the same limiting normal distribution as the optimal M-estimator $\tilde{\underline{\theta}}_n(\gamma_0)$:

$$\sqrt{n}(\tilde{\underline{\theta}}_n(\tilde{\underline{\gamma}}_n) - \theta_0) \to N(0, h(\gamma_0)(\text{plim } \tfrac{1}{n}\underline{X}'\underline{X})^{-1}) \text{ in distr.}, \qquad (6.1.4)$$

provided that ρ is suitable chosen. In the case under review we have chosen

$\rho(u)=\exp(-\frac{1}{2}u^2)/\sqrt{2\pi}$,
which is suitable for (6.1.4).

As recommended in the remark at theorem 3.2.6,before calculating $\tilde{\underline{\gamma}}_n$ we first calculated the γ at which the function $\underline{h}_{2,n}(\gamma,\hat{\underline{\theta}}_n)$ equals zero, where $\hat{\underline{\theta}}_n$ is the L.S.E. Thus $\underline{\theta}_n^*=\hat{\underline{\theta}}_n$ in (3.5.3). We found:

$$\underline{h}_{2,n}(\gamma,\hat{\underline{\theta}}_n)\begin{cases}=0 & \text{at } \gamma=0.0298\ ;\\ <0 & \text{for } \gamma>0.0298\ .\end{cases}$$

Then we plotted the function $\underline{h}_n(\frac{1}{\lambda},\hat{\underline{\theta}}_n)$ on the interval $\lambda\in[0,\frac{1}{\gamma_*}]$, where $\gamma_*=0.05$. The result is shown in figure 6.1.1 below.
From this figure we concluded that the optimal $\tilde{\underline{\gamma}}_n$ is close to 0.1. Notice that the corresponding value of $\underline{h}_n$, i.e. 0.0028, is about 15% lower than the estimated variance of the errors: $\hat{\underline{\sigma}}_n^2=0.0033$ *) [which equals the value of $\underline{h}_n(1/\lambda,\hat{\underline{\theta}}_n)$ at $\lambda=0$].

Next we carried out the robust M-estimation procedure. The iterations were carried out by the modified simplex method of Nelder and Mead (1965), and we used the O.L.S. estimate as initial estimate. Since the robust M-estimator and the least squares estimator are nearly equal for large γ, we started from $\gamma=10$, using the outcome of the iteration as initial estimate for calculation of the M-estimate for the case $\gamma=9.$, and so on until $\gamma=1.$. Then we carried out the same procedure for the cases $\gamma=1.,0.9,0.8,\ldots,0.1$. For the cases $\gamma=10.,9.,\ldots,1.$, and $\gamma=0.9,0.8,0.7$ the M-estimates remained equal to the L.S.E., but from $\gamma=0.6$ we got the following results:

Table 6.1.2 Robust M-estimates

γ	θ_1	θ_2	θ_3	θ_4	θ_5
0.6	-1.623	1.022	0.950	0.004	-8.069
0.5	-1.623	1.022	0.950	0.004	-8.069
0.4	-1.623	1.022	0.950	0.004	-8.069
0.3	-1.627	1.022	0.948	0.004	-8.066
0.2	-1.626	1.022	0.948	0.004	-8.066
0.1	-1.800	1.014	0.981	0.004	-7.986

Using the robust M-estimate $\tilde{\underline{\theta}}_n(.1)$ [the case $\gamma=0.1$], we plotted again the function $\underline{h}_n(\frac{1}{\lambda},\tilde{\underline{\theta}}_n(.1))$ on the interval $[0,\frac{1}{0.05}]$, as is shown in figure 6.1.2 below. We now see that the minimum is still close to $\gamma=0.1$; so we put $\tilde{\underline{\gamma}}_n=0.1$. Moreover, the corresponding value of $\underline{h}_n$ is now about 0.0027.

---

*) $\hat{\underline{\sigma}}_n^2$ is not corrected for loss of degrees of freedom.

FIGURE 6.1.1 : PLOT OF H(GAMMA) BASED ON OLS RESIDUALS

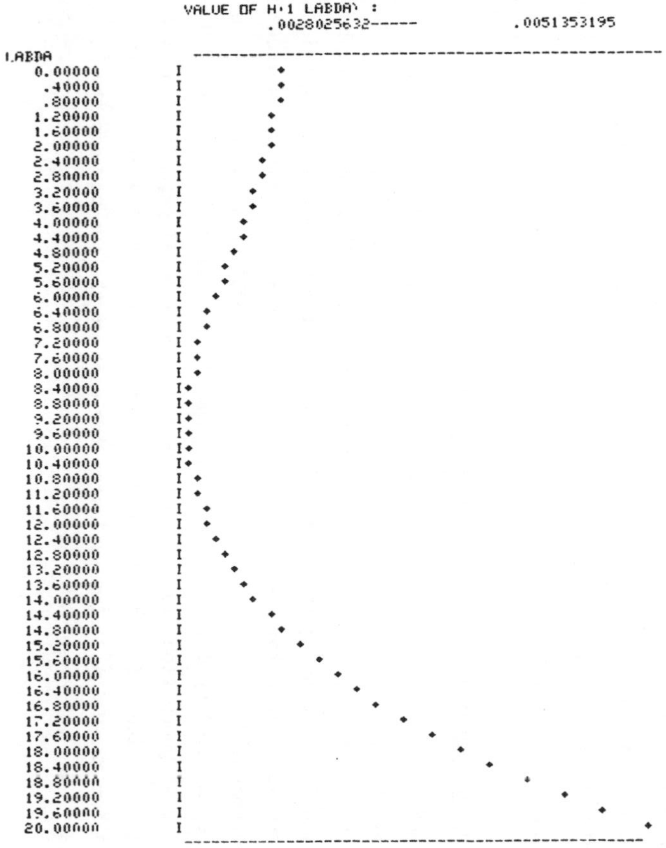

MINIMUM VALUE OF H(1 LABDA) .0028025632
AT LABDA 9.6000000000
CORRESPONDING GAMMA : .1041666667

FIGURE 6.1.2 : PLOT OF H(GAMMA) BASED ON PM RESIDUALS

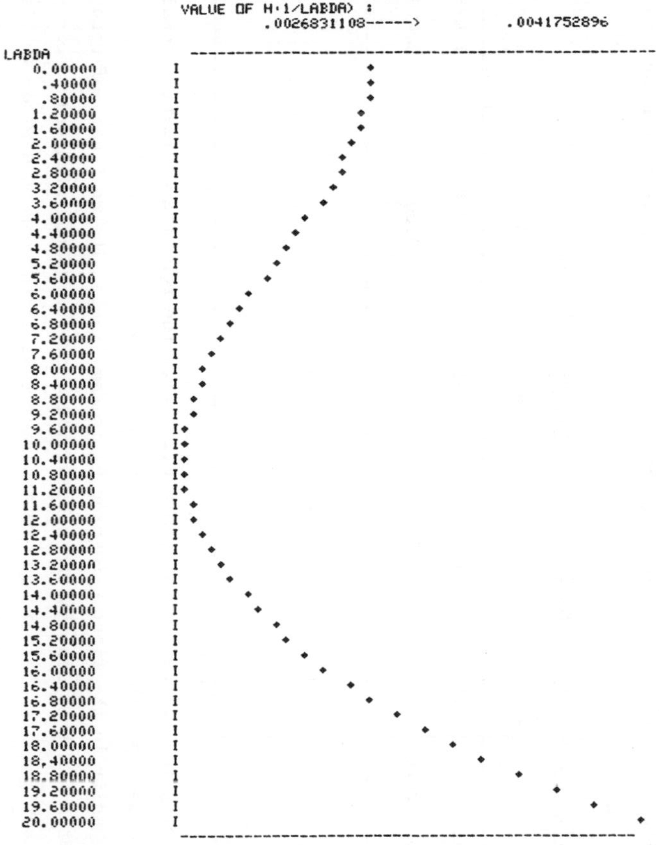

MINIMUM VALUE OF H(1/LABDA) .0026831108
AT LABDA 10.4000000000
CORRESPONDING GAMMA : .0961538462

Finally we estimated the standard errors of the robust M-estimator involved [compare theorem 3.2.7]:

Table 6.1.3 Robust M-estimation results for $\gamma=0.1$

	θ_1	θ_2	θ_3	θ_4	θ_5
coefficient	-1.800	1.014	0.981	0.004	-7.986
standard errors	0.630	0.202	0.155	0.001	2.476
R^2	0.837				
S.E.	0.064	(see table 6.1.1)			
D.W.	1.112				
n	24				

Comparing the OLS results in table 6.1.1 with the robust M-estimation results above, we see that the shifts in the estimated coefficients are not very dramatic; only the shift in the first parameter θ_1 is remarkable. The real gain of robust M-estimation in this case is the substantial decrease of the standard errors of the coefficients. This is due to the fact that in the case under review the variance σ^2 of the errors is about 0.0033 so that by (6.1.1) the limiting normal distribution of the least squares estimator $\hat{\underline{\theta}}_n$ is approximately

$$N(0,0.0033(\text{plim } \tfrac{1}{n}\underline{X}'\underline{X})^{-1}) \quad ,$$

while by (6.1.4) the limiting normal distribution of the two-stage robust M-estimator $\tilde{\underline{\theta}}_n(\tilde{\underline{\gamma}}_n)$ is approximately

$$N(0,0.0027(\text{plim } \tfrac{1}{n}\underline{X}'\underline{X})^{-1}) .$$

6.1.2 An autoregressive model of money demand

The number of observations of the previous example is small, while the Durbin-Watson statistic is rather low. Thus one might be arguing that the observed non-normality of the errors is caused by the small sample size or by autocorrelation, and hence that the example does not provide convincing evidence of non-normality. Therefore we shall consider now a second example of a model for which robust M-estimation is appropriate. This model is a smaller version of the one employed by Sims (1980). Using monthly data of the money stock (M), the industrial production (I), the rate on prime commercial paper (R) and the producers price index for finished goods (P) from 1919-1 to 1941-12 and 1947-1 to 1978-8 for the U.S.A., Sims estimated for each of the two periods the following linear vector autoregression:

$$\underline{y}_t=\sum_{j=1}^{12} B_j\, \underline{y}_{t-j} + c + \underline{u}_t ,$$

where

$$\underline{y}_t=\begin{pmatrix} \ln(\underline{M}_t) \\ \ln(\underline{I}_t) \\ \ln(\underline{R}_t) \\ \ln(\underline{P}_t) \end{pmatrix} ,$$

the $\underline{u}_t$'s are vectors of disturbances, the B_j's are 4x4 matrices of parameters and c is a vector of constants. The purpose of his study is to compare the dynamics of the interwar and the postwar business cycles and to investigate whether the estimated dynamics give support to either the monetarist or the non-monetarist point of view.

For reasons of computational convenience we shall focuss only on the first equation of the shorter model

$$\underline{y}_t=\sum_{j=1}^{3} B_j\, \underline{y}_{t-j} + c + \underline{u}_t \tag{6.1.5}$$

and we shall fit this equation only to the interwar data. Thus the model to be considered is

$$\ln(\underline{M}_t)=\sum_{j=1}^{3}\{\beta_{j,1}\ln(\underline{M}_{t-j})+ \beta_{j,2}\ln(\underline{I}_{t-j})+ \beta_{j,3}\ln(\underline{R}_{t-j})+$$

$$+ \beta_{j,4}\ln(\underline{P}_{t-j})\}+ \beta_5+ \underline{u}_{1,t} \quad , \tag{6.1.6}$$

where $(\beta_{j,1},\beta_{j,2},\beta_{j,3},\beta_{j,4})$ is the upper row of the matrix B_j, β_5 is the first component of the vector c and $\underline{u}_{1,t}$ is the first component of the vector $\underline{u}_t$.

Now assume that the $\underline{u}_t$'s are i.i.d.. If the $\underline{y}_t$'s generated by model (6.1.5) are stationary, then by repeated backwards substitution we can write

$$\underline{y}_t=\sum_{j=0}^{\infty}C_j(\underline{u}_{t-j}+ c)=\sum_{j=0}^{\infty}C_j\underline{u}_{t-j}+ \sum_{j=1}^{\infty}C_jc ,$$

where the C_j's are 4x4 matrices such that the series $\sum_{j=0}^{\infty}C_j$ converges. Putting

$$\underline{y}_t^{(m)} =\sum_{j=0}^{\infty}C_jc + \sum_{j=0}^{m-1}C_j\, \underline{u}_{t-j} ,$$

we then have for fixed t

$$\underline{y}_t^{(m)}\to\underline{y}_t \quad \text{a.s. as } m\to\infty ,$$

hence also

$$\lim_{m\to\infty} P(|\underline{y}_t- \underline{y}_t^{(m)}|>\delta)=0$$

for any $\delta>0$ and consequently by stationarity,

$$\lim_{n\to\infty} \frac{1}{n}\sum_{j=1}^{n} P(|\underline{y}_t - \underline{y}_t^{(m_n)}|>\delta)=0$$

for any sequence (m_n) of positive integers such that $m_n\to\infty$ for $n\to\infty$. So $(\underline{y}_t)$ is in this case stochastically stable with respect to the i.i.d. base $(\underline{u}_t)$. Therefore theorem 5.3.1 is applicable, provided that the distribution of the $\underline{u}_{1,t}$'s is symmetric unimodal.

The estimation of model (6.1.6) is carried out in the same way as for the municipal expenditure model. Thus we first estimated the model by OLS, which yielded the following results:

Table 6.1.4: OLS results

	estimated coefficients	standard errors	statistics
$\beta_{1,1}$	.760	.064	R^2 = .996
$\beta_{1,2}$	.052	.034	S.E. = .014
$\beta_{1,3}$	-.017	.011	D.W. =2.015
$\beta_{2,1}$	.041	.046	n = 273
$\beta_{2,2}$	.288	.075	W.R.K. =6.409
$\beta_{2,3}$	-.035	.056	
$\beta_{3,1}$	.024	.016	(see table 6.1.1)
$\beta_{3,2}$	.106	.068	
$\beta_{3,3}$	-.069	.063	$\hat{\underline{\sigma}}_n^2$=.000177
$\beta_{4,1}$	-.010	.033	(residual variance)
$\beta_{4,2}$	-.017	.011	
$\beta_{4,3}$	-.122	.048	
β_5	.075	.083	

Just as for the example in section 6.1.1 we conclude from these results that the error distribution probably has a finite variance but is leptokurtic, hence non-normal, so that these results may be improved by robust M-estimation. In order to obtain the optimal scaling parameter γ we estimated the function $h(\gamma)$ by $\underline{h}_n(\gamma,\hat{\underline{\theta}}_n)$ [see section 3.2.5], where $\hat{\underline{\theta}}_n$ is the vector of OLS estimates, and we plotted the function $\underline{h}_n(\frac{1}{\lambda},\hat{\underline{\theta}}_n)$ on the interval [0,49], as is shown in figure 6.1.3. From this figure we concluded that the optimal γ is approximately $\frac{1}{31}$. Next we re-estimated the model, using robust M-estimation with the same ρ as in section 6.1.1. As reported in table 6.1.2, we started with a large γ, using the OLS estimates as initial estimates, and then we repeated the procedure for a finite, monotonically decreasing sequence of γ's, the last of

FIGURE 6.1.3 : PLOT OF H(GAMMA) BASED ON OLS RESIDUALS

VALUE OF H(1/LABDA) :

.0001308491-----> .0001765137

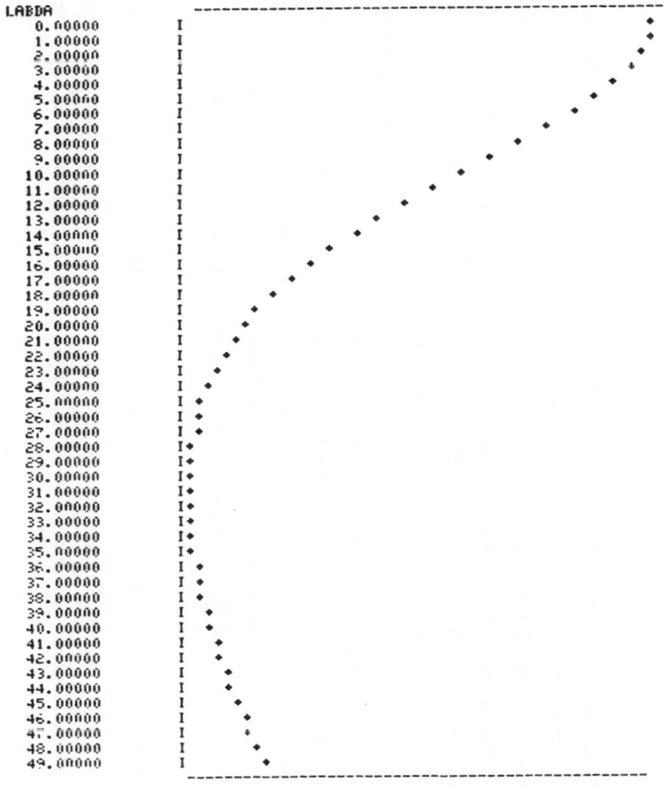

MINIMUM VALUE OF H(1/LABDA) .0001308491
AT LABDA 31.0000000000
CORRESPONDING GAMMA : .0322580645

FIGURE 6.1.4 : PLOT OF H(GAMMA) BASED ON RM RESIDUALS

VALUE OF H(1/LABDA) :

.0001271447-----> .0001801679

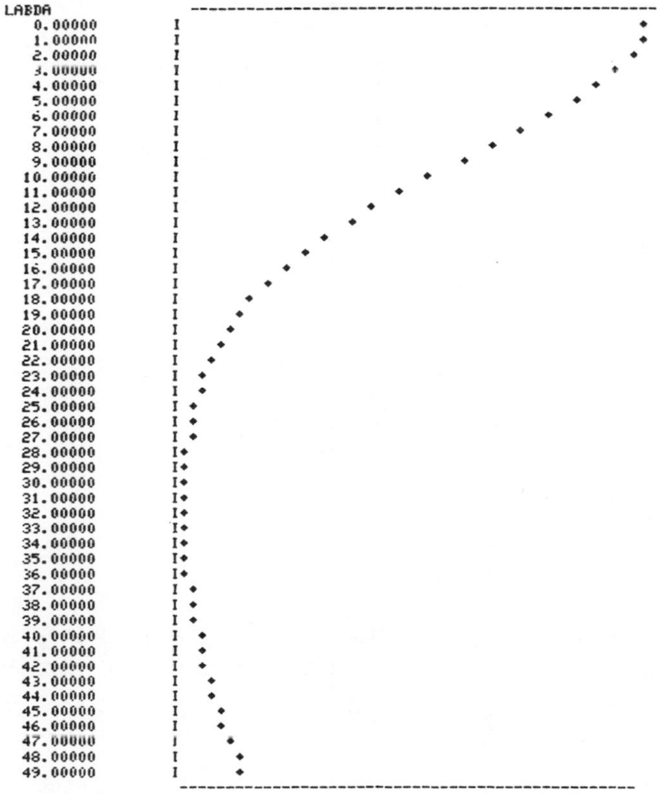

MINIMUM VALUE OF H(1/LABDA) .0001271447
AT LABDA 31.0000000000
CORRESPONDING GAMMA : .0322580645

these γ's being $\gamma = \frac{1}{31}$. On the basis of the robust M-estimation residuals for $\gamma = \frac{1}{31}$ we once more plotted the function $\underline{h}_n$, as is shown in figure 6.1.4, which shows that the optimal γ is still close to $\frac{1}{31}$. Therefore we have set

$$\tilde{\underline{\gamma}}_n = \frac{1}{31}.$$

The robust M-estimation results for this γ are reported in table 6.1.5 below.

Table 6.1.5: Robust M-estimation results for $\gamma = \frac{1}{31}$

	estimated coefficients	standard errors	statistics	
$\beta_{1,1}$	.904	.054	R^2	= .996
$\beta_{1,2}$	.026	.031	S.E.	= .014
$\beta_{1,3}$	-.010	.009	D.W.	=2.265
$\beta_{2,1}$	.027	.039	n	= 273
$\beta_{2,2}$	.210	.064	$\underline{h}_n$	= .000127
$\beta_{2,3}$	-.014	.046	$\hat{\underline{\sigma}}_n^2$	= .000180
$\beta_{3,1}$	.020	.014		
$\beta_{3,2}$	.125	.057		
$\beta_{3,3}$	-.126	.053		
$\beta_{4,1}$	-.010	.030		
$\beta_{4,2}$	-.017	.009		
$\beta_{4,3}$	-.136	.040		
β_5	.053	.071		

Comparing the results in table 6.1.4 with those in table 6.1.5, we see a substantial upwards shift of the estimate of $\beta_{1,1}$, the coefficient of the one month lagged log of the money stock. This is especially remarkable in view of 95% confidence intervals of $\beta_{1,1}$, i.e. [.635, .885] on basis of the OLS results and [.798, 1.010] on basis of the robust M-estimation results, so that the robust M-estimate lies outside the 95% confidence interval based on OLS estimation, and so does the OLS estimate with respect to the 95% confidence interval based on robust M-estimation. However, these confidence intervals are only approximations. The error distribution is strongly leptokurtic, as follows from the high value of the statistic WRK, i.e. 6.409, hence its tails are much fatter than in the case of a normal distribution. Therefore the true 95% confidence interval of $\beta_{1,1}$, especially the one based on OLS results, may be larger.

Also other estimates are shifted, and some of them are now becoming significant.

We recall that the main purpose of this example is to show that non-normal error distributions exist and that robust M-estimation makes sense. The non-normality clearly follows from the high value of the statistic WRK, namely 6.409, while in the normal case we should find a WRK close to zero. Moreover, the non-normality of the errors is confirmed by the figures 6.1.3 and 6.1.4, because in the normal case the function $h(\gamma)$ cannot take values less than σ^2, the variance of the error distribution, because then least squares estimators equal maximum likelihood estimators, while maximum likelihood estimators are known to be asymptotically most efficient under some conditions. For the OLS estimator $\hat{\underline{\theta}}_n$ satisfies (6.1.1):

$$\sqrt{n}(\hat{\underline{\theta}}_n - \theta_0) \rightarrow N[0,\sigma^2(\text{plim } \tfrac{1}{n}\,\underline{X}'\underline{X})^{-1}] \quad \text{in distr.},$$

where now the matrix $\underline{X}$ consists of the explanatory variables of model (6.1.6), while the robust M-estimator $\tilde{\underline{\theta}}_n(\gamma)$ satisfies

$$\sqrt{n}(\tilde{\underline{\theta}}_n(\gamma) - \theta_0) \rightarrow N[0,h(\gamma)(\text{plim } \tfrac{1}{n}\underline{X}'\underline{X})^{-1}] \text{ in distr.}$$

If the error distribution is normal, then the least squares estimator equals the maximum likelihood estimator, hence there cannot be another estimator with a lower asymptotic variance. Thus in the normal case $h(\gamma)$ cannot be lower than σ^2. But the minimum value of the estimator of $h(\gamma)$ is .000127 at $\gamma = \frac{1}{31}$ while the estimate of σ^2 is .000180, which contradicts the normality hypothesis.

6.2 An application of minimum information estimation

The model considered here relates the money stock, as a percentage of net national income, to the money market rate. But there is no one side causality, because money supply and demand obviously also may affect the money market rate. Such a relationship should therefore be considered as a structural equation of a larger, but further unspecified, simultaneous equation system.

Denoting the money stock, relative to net national income, by $\underline{M}$ and the money market rate by $\underline{R}$, we specified the model as

$$\ln(\underline{M}) = \alpha . \ln(\underline{R}) + \beta + \underline{v} ,$$

where $\underline{v}$ is a disturbance term. However, using Dutch annual data from 1948 to 1973, preliminary OLS regression results suggested autocorrelation of the disturbances:

$$\underline{v}_t = \rho \underline{v}_{t-1} + \underline{u}_t .$$

Therefore we restated the model as

$$\ln(\underline{M}_t) = \rho . \ln(\underline{M}_{t-1}) + \alpha\{\ln(\underline{R}_t) - \rho . \ln(\underline{R}_{t-1})\} + \beta(1-\rho) + \underline{u}_t . \qquad (6.2.1)$$

For applying the minimum information estimation (MIE) method, it is desirable to use an instrumental variable. We have chosen $\ln(\underline{D}_t)$, where $\underline{D}_t$ is the discount rate of the Dutch Central Bank, as the instrumental variable, because this variable clearly has an impact on both the money supply and the money market rate, but should be considered as exogenous, so that $\underline{u}_t$ and $\ln(\underline{D}_t)$ may be assumed to be mutually independent.

Putting

$$\underline{z}_t = \begin{pmatrix} \ln(\underline{M}_t) \\ \ln(\underline{M}_{t-1}) \\ \ln(\underline{R}_t) \\ \ln(\underline{R}_{t-1}) \\ \ln(\underline{D}_t) \end{pmatrix}, \quad \theta_0 = \begin{pmatrix} \rho \\ \alpha \\ \beta(1-\rho) \end{pmatrix},$$

model (6.2.1) can be written as a nonlinear implicit structural equation

$$r(\underline{z}_t,\theta_0)=\underline{u}_t ,$$

where $r(z,\theta)=z_1-\theta_1 z_2-\theta_2(z_3-\theta_1 z_4)-\theta_3$ with $z'=(z_1,z_2,z_3,z_4,z_5)$. The instrumental variable is now [compare assumption 4.3.1]

$$w(\underline{z}_t)=\ln(\underline{D}_t).$$

If the $\underline{u}_t$'s are i.i.d. with common symmetric distribution, and if the $\underline{z}_t$'s are stochastically stable with respect to an independent base, then the results of section 5.3.2 indicates that the MIE-method is appropriate for the model under review.

In order to obtain an initial estimate of θ_0, we estimated model (6.2.1) by OLS, which yielded the following result:

$$\ln(\underline{M}_t)=.697\ln(\underline{M}_{t-1})-.101\ \ln(\underline{R}_t)+\ 0.056\ln(\underline{R}_{t-1})+1.058$$

R^2 = .956

S.E. = .040

D.W. =2.267

n = 25 .

This suggests

$$\rho \approx .7 \ , \ \alpha \approx -.1 \ , \ \beta(1-\rho) \approx 1. \quad .$$

Using these values as an initial estimate of θ_0, and chosing the gamma density (4.4.3) with k=2 as weight function, we carried out the estimation as follows.

For $\gamma=\gamma_1=1.$ we made four plots of the objective function $\underline{S}_n^*(\theta|\gamma_1)$ *), as is shown in the figures 6.2.1/1 through 6.2.1/4. These plots show the shape of the

*) See (4.4.1).

objective function for

$$\underline{\theta} = \theta_1 + \delta . \frac{\underline{\xi}}{|\underline{\xi}|} \qquad (6.2.3)$$

where θ_1 is the initial estimate $[\theta_1'=(.7,-.1,1.)]$, each component of the vector $\underline{\xi}'=(\underline{\xi}_1,\underline{\xi}_2,\underline{\xi}_3)$ is a random drawing from the uniform [-1,1] distribution and δ runs through the interval [-.5,+.5]. Thus the plots show the shape of $\underline{S}_n^*(\theta|\gamma_1)$ on the line between the random points $\theta_1-.5\frac{\underline{\xi}}{|\underline{\xi}|}$ and $\theta_1+.5\frac{\underline{\xi}}{|\underline{\xi}|}$.

We see from these figures that the initial estimate $\theta_1'=(.7,-.1,1.)$ [point 15], lies in the crater. Thus it looks safe to carry out the iteration [using the modified simplex method of Nelder and Mead (1965)]. Doing so, we obtained the following estimate:

$$\underline{\theta}_n^o(\gamma_1)'=(.709,-.103,1.001),$$

where $\gamma_1=1$.

We recall (see theorems 4.3.2 and 5.3.4) that for every $\gamma>0$ we have

$$\sqrt{n}\ (\underline{\theta}_n^o(\gamma)-\theta_0)\to N[0,\Sigma_2(\theta_0|\gamma)^{-1}\Sigma_1(\theta_0|\gamma)\Sigma_2(\theta_0|\gamma)^{-1}] \quad \text{in distr.}.$$

Moreover, from theorem 4.4.4 it follows that the asymptotic variance matrix involved is bounded from above by $\Sigma_2(\theta_0|\gamma)^{-1}$, which in its turn can be estimated consistently by the matrix $\underline{\Sigma}_{2,n}(\underline{\theta}_n^o(\gamma)|\gamma)^{-1}$ defined by (4.3.41). Consequently, the square roots of the diagonal elements of the matrix $\frac{1}{n}\underline{\Sigma}_{2,n}(\underline{\theta}_n^o(\gamma)|\gamma)^{-1}$ may be used as upperbounds of the standard errors of the parameter estimates. We have calculated these upperbounds instead of the standard errors themselves, in order to save computer time.

Table 6.2.1: MIE-results for the case $\gamma=1$

	ρ	α	$\beta(1-\rho)$
$\underline{\theta}_n^o(\gamma)$	.709	-.103	1.001
upperbounds of the standard errors	2225.991	2562.429	8821.889
minimum value of the objective function	.0000008751		
n	25		

Thus for the case $\gamma=1$ the estimates are extremely insignificant. This indicates that the scaling parameter γ is too low. Therefore we have plotted the function

$$\underline{\overline{T}}_n^*(\gamma|1.)=\text{tr}.[\underline{\Sigma}_{2,n}(\underline{\theta}_n^o(1.)|\gamma)^{-1}] \qquad (6.2.4)$$

(see section 4.4.3) on the interval $\Gamma=[1,30]$, as is shown in the figures 6.2.2/1 through 6.2.2/3. These plots show that $\underline{\overline{T}}_n^*(\gamma|1)$ is minimal on Γ at γ=30.

In view of the argument in the sections 4.3.5 and 4.4.3 we repeated the estimation procedure for the case γ=30,using (6.2.2) as initial estimate. But before carrying out the iteration, we verified whether this initial estimate lies in, or on the wall of, the crater. Thus again we made four plots in random directions, similar to the figures 6.2.1/1 through 6.2.1/4, but now for γ=30 and the initial estimate (6.2.2). Moreover, now δ in (6.2.3) is running through the interval [-.1,.1] instead of [-.5,.5] in the case of the figures 6.2.1/1 through 6.2.1/4. The results are shown in the figures 6.2.3/1 through 6.2.3/4. Also now the initial estimate[point 15]is lying in the crater. We see that in this case the crater is narrower than in the case shown in the figures 6.2.1/1 through 6.2.1/4. This is due to the higher value of γ. Carrying out the iteration, and estimating the upperbounds of the standard errors of the parameter estimates as before, we obtained the following results:

Table 6.2.2: MIE-results for the case γ=30

	ρ	α	$\beta(1-\rho)$
$\underline{\theta}_n^o(\gamma)$	.695	-.115	1.042
upperbounds of the standard errors	.218	.053	.779
minimum value of the objective function	.004381		
S.E.	.0407		
n	25		

The parameter estimates in table 6.2.2 do not differ very much from the results obtainable from applying nonlinear least squares, as can be observed from table 6.2.3 below:

Table 6.2.3: Nonlinear least squares results

	ρ	α	$\beta(1-\rho)$
$\hat{\underline{\theta}}_n$	.737	-.102	.904
standard errors	.056	.020	.200
S.E.	.0406		
R^2	.953		
n	25		

FIGURE 6.2.1-1

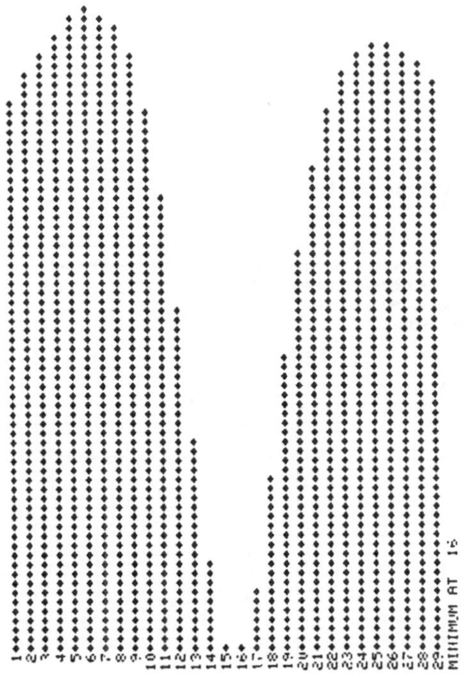

FIGURE 6.2.1 2

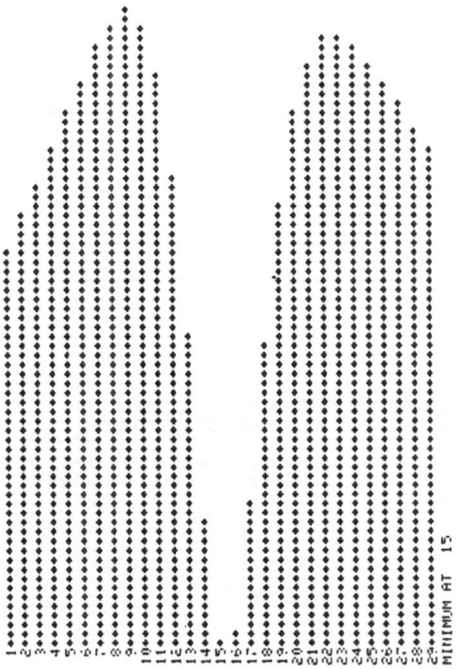

FIGURE 6.2.1 3

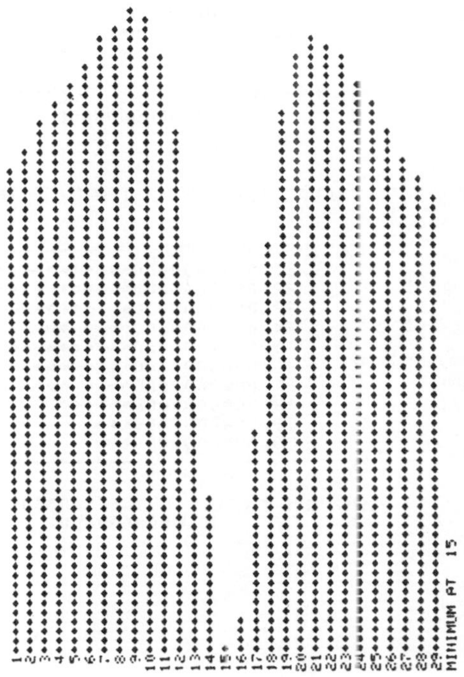

FIGURE 6.2.1 4

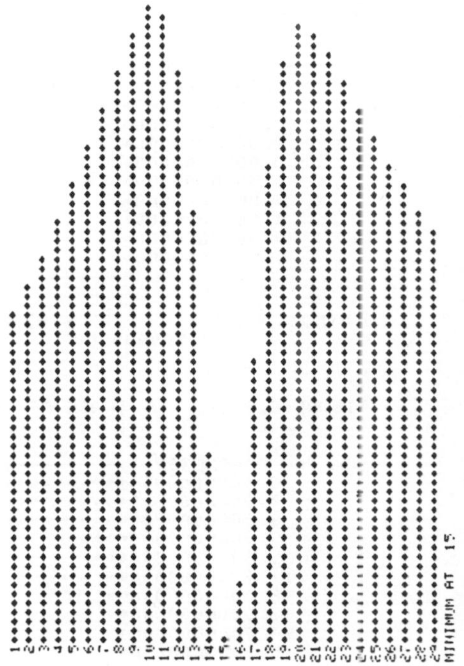

```
FIGURE 6.2.2/1      TRACE OF THE UPPERBOUND OF THE VARIANCE MATRIX:
                          177.6398722851-----> 89346811.2058157921

SCALING PARAMETER  --------------------------------------------------
       1.0000000000◆◆◆◆◆◆◆◆◆◆◆◆◆◆◆◆◆◆◆◆◆◆◆◆◆◆◆◆◆◆◆◆◆◆◆◆◆◆◆◆◆◆◆◆◆◆◆◆◆◆
       1.5000000000◆◆◆
       2.0000000000◆
       2.5000000000◆
       3.0000000000◆
       3.5000000000◆
       4.0000000000◆
       4.5000000000◆
       5.0000000000◆
       5.5000000000◆
                   --------------------------------------------------

MINIMUM TRACE         177.6398722851
AT            5.5000000000
```

```
FIGURE 6.2.2/2      TRACE OF THE UPPERBOUND OF THE VARIANCE MATRIX:
                            .2037305615------>        94.3505308981

SCALING PARAMETER  --------------------------------------------------
       6.0000000000◆◆◆◆◆◆◆◆◆◆◆◆◆◆◆◆◆◆◆◆◆◆◆◆◆◆◆◆◆◆◆◆◆◆◆◆◆◆◆◆◆◆◆◆◆◆◆◆◆◆
       7.0000000000◆◆◆◆◆◆◆◆◆◆◆◆◆◆◆◆◆◆◆◆
       8.0000000000◆◆◆◆◆◆◆◆◆◆
       9.0000000000◆◆◆◆◆◆
      10.0000000000◆◆◆
      11.0000000000◆◆
      12.0000000000◆◆
      13.0000000000◆◆
      14.0000000000◆
      15.0000000000◆
      16.0000000000◆
      17.0000000000◆
      18.0000000000◆
      19.0000000000◆
      20.0000000000◆
                   --------------------------------------------------

MINIMUM TRACE             .2037305615
AT            20.0000000000
```

```
FIGURE 6.2.2/3      TRACE OF THE UPPERBOUND OF THE VARIANCE MATRIX:
                            .0750500428------>          .1746661877

SCALING PARAMETER  --------------------------------------------------
      21.0000000000◆◆◆◆◆◆◆◆◆◆◆◆◆◆◆◆◆◆◆◆◆◆◆◆◆◆◆◆◆◆◆◆◆◆◆◆◆◆◆◆◆◆◆◆◆◆◆◆◆◆
      22.0000000000◆◆◆◆◆◆◆◆◆◆◆◆◆◆◆◆◆◆◆◆◆◆◆◆◆◆◆◆◆◆◆◆◆◆◆◆◆◆◆
      23.0000000000◆◆◆◆◆◆◆◆◆◆◆◆◆◆◆◆◆◆◆◆◆◆◆◆◆◆◆◆◆◆
      24.0000000000◆◆◆◆◆◆◆◆◆◆◆◆◆◆◆◆◆◆◆◆◆◆◆
      25.0000000000◆◆◆◆◆◆◆◆◆◆◆◆◆◆◆◆◆
      26.0000000000◆◆◆◆◆◆◆◆◆◆◆◆◆
      27.0000000000◆◆◆◆◆◆◆◆◆
      28.0000000000◆◆◆◆◆◆
      29.0000000000◆◆◆
      30.0000000000◆
                   --------------------------------------------------

MINIMUM TRACE            .0750500428
AT            30.0000000000
```

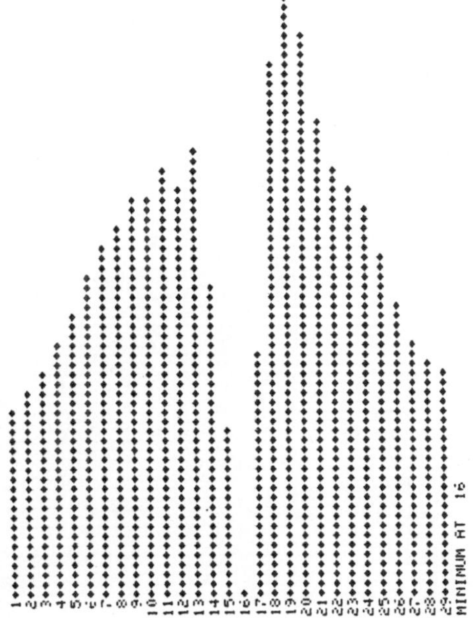

FIGURE 6.2.3.1

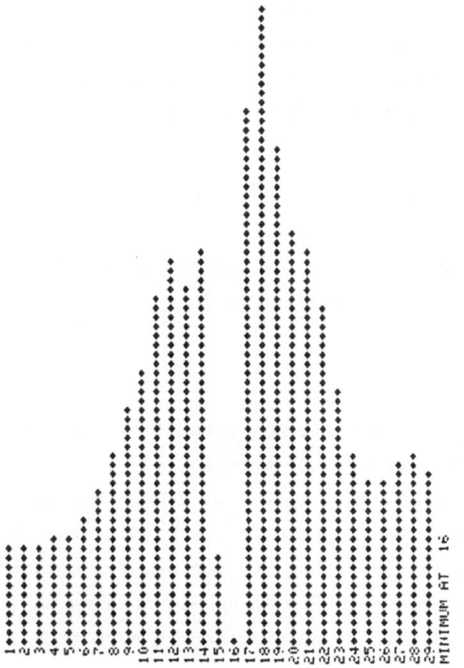

FIGURE 6.2.3 2

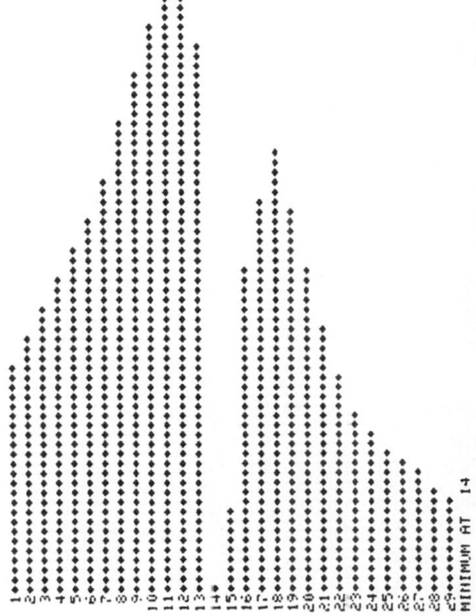

FIGURE 6.2.3 3

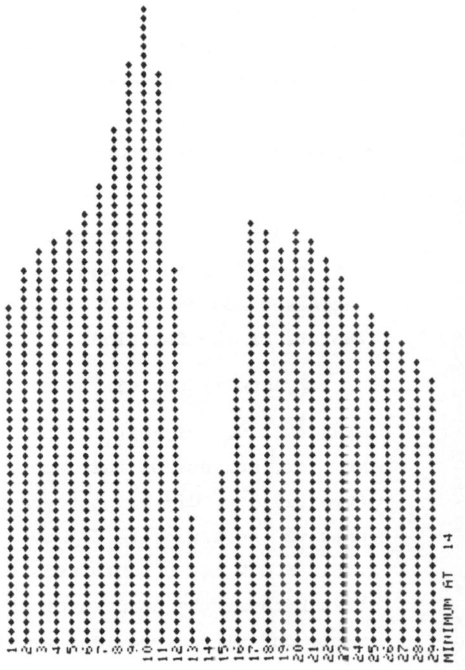

FIGURE 6.2.3 4

Possibly this is due to the small sample size, by which the data does not contain enough information. Nevertheless, from an asymptotic point of view nonlinear least squares is not an appropriate estimation method for this model, because we may expect that the explanatory variables are more or less dependent of the errors.

We have seen from the figures 6.2.2/1 through 6.2.2/3 that the function $\underline{\overline{T}}_n^*(\gamma|1.)$ is strongly decreasing on $\Gamma=[1,30]$. We have done some other experiments with our MIE-method, and in all cases we found that by increasing γ this function decreases monotonically to a small but nonzero minimum. Possibly this is a feature of the function

$$\overline{T}^*(\gamma)=\text{tr.}[\Sigma_2(\theta_0|\gamma)^{-1}]$$

[(see (4.4.22)] itself, of which $\underline{\overline{T}}_n^*(\gamma|\gamma_1)$ is a uniformly consistent estimator. However, because of the complicated structure of the matrix $\Sigma_2(\theta_0|\gamma)$ [see (4.3.31)], we did not succeed in proving it. But if this is true, then MIE-results can always be improved by increasing γ; hence better results than in table 6.3.2 may be obtained by using a larger γ than 30. But we have not done this because the purpose of the example under review is merely to show how to apply our MIE-method, and not to make a serious contribution to economics itself.

REFERENCES

AMEMIYA, T. (1974), The Nonlinear Two Stage Least Squares Estimator, *Journal of Econometrics*, 2, p. 105-110.

AMEMIYA, T. (1975), The Nonlinear Limited Information Maximum Likelihood Estimator and the Modified Two Stage Least Squares Estimator, *Journal of Econometrics*, 3, p. 375-386.

ANDERSON, T.W. (1958), *An Introduction to Multivariate Statistical Analysis*, New York-London: John Wiley and Sons, Inc.

ANDREWS, D.F., P.J. BICKEL, F.R. HAMPEL, P.J. HUBER, W.H. ROGERS and J.W. TUCKEY (1972), *Robust Estimates of Location: Survey and Advances*, Princeton: Princeton University Press.

ANDREWS, D.F. (1974), A Robust Method for Multiple Linear Regression, *Technometrics*, 16, p. 523-531.

BERNDT, E.K., R.E. HALL and J.A. HAUSMAN, (1974), Estimation and Inference in Nonlinear Structural Models, *Annals of Economic and Social Measurement* 3/4, p. 653-665.

BIERENS, H.J. (1980), Consistent Selection of Explanatory Variables, *Statistica Neerlandica* 34, p. 141-150.

BIERENS, H.J., J.D. POELERT and R.A. STOFFEL (1980), *Vergelijkend onderzoek naar de financiële structuur en het voorzieningenniveau van Amsterdam, Rotterdam, Den Haag en Utrecht; Deel 2: Structuur van uitgaven en inkomsten over de periode 1948-1974*, Amsterdam: Stichting voor Economisch Onderzoek der Universiteit van Amsterdam.

BILLINGSLEY, P. (1968), *Convergence of Probability Measures*, New York-London: John Wiley and Sons, Inc.

BOX, G.E.P. and D.R. COX (1964), An Analysis of Transformations, *Journal of the Royal Statistical Society*, Series B, 26, p. 211-243.

BROWN, B.M. (1971), Martingale Central Limit Theorems, *The Annals of Mathematical Statistics*, 42, p. 59-66.

CHUNG, K.L. (1974), *A Course in Probability Theory*, New York-London: Academic Press.

CRAMER, J.S. (1969), *Empirical Econometrics*, Amsterdam: North-Holland Publishing Co.

EISENPRESS, H. and J.G. GREENSTADT (1966), The Estimation of Nonlinear Econometric Systems, *Econometrica*, 34, p. 851-861.

FAMA, E.F. (1963), Mandelbrot and the Stable Paretian Hypothesis, *Journal of Business*, 36, p. 420-429.

FAMA, E.F. (1965), The Behaviour of Stock-market Prices, *Journal of Business*, 38, p. 34-105.

FELLER, W. (1966), *An Introduction to Probability Theory and Its Applications*, Vol. II, New York-London: John Wiley and Sons, Inc.

FORSYTHE, A.B. (1972), Robust Estimation of Straight Line Regression Coefficients by Minimizing p-th Power Deviations, *Technometrics*, 14, p. 159-166.

GALLANT, A.R. (1977), Three Stage Least Squares Estimation for a System of Simultaneous, Nonlinear, Implicit Equations, *Journal of Econometrics*, 5, p. 70-88.

GIHMAN, I.I. and A.V. SKOROHOD (1974), *The Theory of Stochastic Processes I*, Berlin: Springer-Verlag.

GNEDENKO, B.V. and A.N. KOLMOGOROV (1954), *Limit Distribution for Sums of Independent Random Variables*, Reading, Mass.: Addison-Wesley Publishing Co.

GOLDFELD, S.M. and R.E. QUANDT (1972), *Nonlinear Methods in Econometrics*, Amsterdam: North-Holland Publishing Co.

HAMPEL, F.R. (1971), A General Qualitative Definition of Robustness, *The Annals of Mathematical Statistics*, 42, p. 1887-1896.

HAMPEL, F.R. (1974), The Influence Curve and Its Role in Robust Estimation, *Journal of the American Statistical Association*, 69, p. 383-393.

HANNAN, E.J. and M. KANTER (1974), Autoregressive Processes with Infinite Variance, *Journal of Applied Probability*, 14, p. 411-415.

HARTLEY, H.O. and A. BROOKER (1965), Nonlinear Least Squares Estimation, *The Annals of Mathematical Statistics*, 36, p. 638-650.

HUBER, P.J. (1964), Robust Estimation of A Location Parameter, *The Annals of Mathematical Statistics*, 35, p. 73-101.

HUBER, P.J. (1972), Robust Statistics, *The Annals of Mathematical Statistics*, 43, p. 1041-1067.

HUBER, P.J. (1973), Robust Regression: Asymptotics, Conjecture and Monte Carlo, *Annals of Statistics*, 1, p. 799-821.

JENNRICH, R.I. (1969), Asymptotic Properties of Nonlinear Least Squares Estimators, *The Annals of Mathematical Statistics*, 40, p. 633-643.

JORGENSON, D.W. and J.J. LAFFONT (1974), Efficient Estimation of Nonlinear Simultaneous Equations with Additive Disturbances, *Annals of Economic and Social Measurement*, 3/4, p. 615-639.

KELEJIAN, H.H. (1972), Two Stage Least Squares and Economic Systems Linear in the Parameters but Nonlinear in Endogenous Variables, *Journal of the American Statistical Association*, 66, p. 373-374.

KOLMOGOROV, A.N. and S.V. FOMIN (1957), *Elements of the Theory of Functions and Functional Analysis, Vol.1: Metric and Normed Spaces*, Rochester, New York: Graylock Press.

KOLMOGOROV, A.N. and S.V. FOMIN (1961), *Elements of the Theory of Functions and Functional Analysis, Vol.2: Measure. The Lebesgue Integral. Hilbert Space*, Rochester, New York: Graylock Press.

LEECH, D. (1975), Testing the Error Specification in Nonlinear Regression, *Econometrica*, 43, p. 719-725.

LUKACS, E. (1970), *Characteristics Functions*, London: Griffin.

MALINVAUD, E. (1970), *Statistical Methods of Econometrics*, Amsterdam: North-Holland Publishing Co.

MALINVAUD, E. (1970), The Consistency of Nonlinear Regression, *The Annals of Mathematical Statistics*, 41, p. 956-969.

MANDELBROT, B. (1963), New Methods in Statistical Economics, *The Journal of Political Economy*, 71, p. 421-440.

MANDELBROT, B. (1963), The Variation of Certain Speculative Prices, *Journal of Business*, 36, p. 394-419.

MANDELBROT, B. (1967), The Variation of Some other Speculative Prices, *Journal of Business*, 40, p. 393-413.

NELDER, J.A. and R. MEAD (1965), A Simplex Method for Function Minimization, *The Computer Journal*, 7, p. 308-313.

REVESZ, P. (1968), *The Laws of Large Numbers*, New York-London: Academic Press.

RONNER, A.E. (1977), *P-Norm Estimators in a Linear Regression Model*, Ph.D. Thesis, University of Groningen.

ROYDEN, H.L. (1968), *Real Analysis*, London: Macmillan Co.

RUDIN, W. (1976), *Principals of Mathematical Analysis*, Tokyo: Mac Graw-Hill Kogakusha, Ltd.

SIMS, C.A. (1976), *Asymptotic Distribution Theory for a Class of Nonlinear Estimation Methods*, Discussion Paper No. 76-69, Centre for Economic Research, Department of Economics, University of Minnesota.

SIMS, C.A. (1980), *Comparison of Interwar and Postwar Cycles: Monetarism Reconsidered,* Working Paper No. 430, Cambridge: National Bureau of Economic Research.

SPITZER, J.J. (1976), The Demand for Money, the Liquidity Trap, and Functional Forms, *International Economic Review,* 17, p. 220-227.

STOUT, W.F. (1974), *Almost Sure Convergence,* New York-London: Academic Press.

THEIL, H. (1971), *Principles of Econometrics,* Amsterdam: North-Holland Publishing Co.

VILLEGAS, C. (1969), On the Least Squares Estimation of Nonlinear Relations, *The Annals of Mathematical Statistics,* 40, p. 426-446.

WHITE, K.J. (1972), Estimation of the Liquidity Trap with a Generalized Functional Form, *Econometrica,* 40, p. 193-199.

WILKS, S.S. (1963), *Mathematical Statistics,* Princeton: Princeton University Press.

WITTING, H. and G. NÖLLE (1970), *Angewandte Mathematische Statistik,* Stuttgart: B.G. Teubner.

ZAREMBKA, P. (1968), Functional Form in the Demand for Money, *Journal of the American Statistical Association,* 18, p. 502-511.

Vol. 83: NTG/GI-Gesellschaft für Informatik, Nachrichtentechnische Gesellschaft. Fachtagung „Cognitive Verfahren und Systeme", Hamburg, 11.–13. April 1973. Herausgegeben im Auftrag der NTG/GI von Th. Einsele, W. Giloi und H.-H. Nagel. VIII, 373 Seiten. 1973.

Vol. 84: A. V. Balakrishnan, Stochastic Differential Systems I. Filtering and Control. A Function Space Approach. V, 252 pages. 1973.

Vol. 85: T. Page, Economics of Involuntary Transfers: A Unified Approach to Pollution and Congestion Externalities. XI, 159 pages. 1973.

Vol. 86: Symposium on the Theory of Scheduling and its Applications. Edited by S. E. Elmaghraby. VIII, 437 pages. 1973.

Vol. 87: G. F. Newell, Approximate Stochastic Behavior of n-Server Service Systems with Large n. VII, 118 pages. 1973.

Vol. 88: H. Steckhan, Güterströme in Netzen. VII, 134 Seiten. 1973.

Vol. 89: J. P. Wallace and A. Sherret, Estimation of Product. Attributes and Their Importances. V, 94 pages. 1973.

Vol. 90: J.-F. Richard, Posterior and Predictive Densities for Simultaneous Equation Models. VI, 226 pages. 1973.

Vol. 91: Th. Marschak and R. Selten, General Equilibrium with Price-Making Firms. XI, 246 pages. 1974.

Vol. 92: E. Dierker, Topological Methods in Walrasian Economics. IV, 130 pages. 1974.

Vol. 93: 4th IFAC/IFIP International Conference on Digital Computer Applications to Process Control, Part I. Zürich/Switzerland, March 19–22, 1974. Edited by M. Mansour and W. Schaufelberger. XVIII, 544 pages. 1974.

Vol. 94: 4th IFAC/IFIP International Conference on Digital Computer Applications to Process Control, Part II. Zürich/Switzerland, March 19–22, 1974. Edited by M. Mansour and W. Schaufelberger. XVIII, 546 pages. 1974.

Vol. 95: M. Zeleny, Linear Multiobjective Programming. X, 220 pages. 1974.

Vol. 96: O. Moeschlin, Zur Theorie von Neumannscher Wachstumsmodelle. XI, 115 Seiten. 1974.

Vol. 97: G. Schmidt, Über die Stabilität des einfachen Bedienungskanals. VII, 147 Seiten. 1974.

Vol. 98: Mathematical Methods in Queueing Theory. Proceedings 1973. Edited by A. B. Clarke. VII, 374 pages. 1974.

Vol. 99: Production Theory. Edited by W. Eichhorn, R. Henn, O. Opitz, and R. W. Shephard. VIII, 386 pages. 1974.

Vol. 100: B. S. Duran and P. L. Odell, Cluster Analysis. A Survey. VI, 137 pages. 1974.

Vol. 101: W. M. Wonham, Linear Multivariable Control. A Geometric Approach. X, 344 pages. 1974.

Vol. 102: Analyse Convexe et Ses Applications. Comptes Rendus, Janvier 1974. Edited by J.-P. Aubin. IV, 244 pages. 1974.

Vol. 103: D. E. Boyce, A. Farhi, R. Weischedel, Optimal Subset Selection. Multiple Regression, Interdependence and Optimal Network Algorithms. XIII, 187 pages. 1974.

Vol. 104: S. Fujino, A Neo-Keynesian Theory of Inflation and Economic Growth. V, 96 pages. 1974.

Vol. 105: Optimal Control Theory and its Applications. Part I. Proceedings 1973. Edited by B. J. Kirby. VI, 425 pages. 1974.

Vol. 106: Optimal Control Theory and its Applications. Part II. Proceedings 1973. Edited by B. J. Kirby. VI, 403 pages. 1974.

Vol. 107: Control Theory, Numerical Methods and Computer Systems Modeling. International Symposium, Rocquencourt, June 17–21, 1974. Edited by A. Bensoussan and J. L. Lions. VIII, 757 pages. 1975.

Vol. 108: F. Bauer et al., Supercritical Wing Sections II. A Handbook. V, 296 pages. 1975.

Vol. 109: R. von Randow, Introduction to the Theory of Matroids. IX, 102 pages. 1975.

Vol. 110: C. Striebel, Optimal Control of Discrete Time Stochastic Systems. III. 208 pages. 1975.

Vol. 111: Variable Structure Systems with Application to Economics and Biology. Proceedings 1974. Edited by A. Ruberti and R. R. Mohler. VI, 321 pages. 1975.

Vol. 112: J. Wilhelm, Objectives and Multi-Objective Decision Making Under Uncertainty. IV, 111 pages. 1975.

Vol. 113: G. A. Aschinger, Stabilitätsaussagen über Klassen von Matrizen mit verschwindenden Zeilensummen. V, 102 Seiten. 1975.

Vol. 114: G. Uebe, Produktionstheorie. XVII, 301 Seiten. 1976.

Vol. 115: Anderson et al., Foundations of System Theory: Finitary and Infinitary Conditions. VII, 93 pages. 1976

Vol. 116: K. Miyazawa, Input-Output Analysis and the Structure of Income Distribution. IX, 135 pages. 1976.

Vol. 117: Optimization and Operations Research. Proceedings 1975. Edited by W. Oettli and K. Ritter. IV, 316 pages. 1976.

Vol. 118: Traffic Equilibrium Methods, Proceedings 1974. Edited by M. A. Florian. XXIII, 432 pages. 1976.

Vol. 119: Inflation in Small Countries. Proceedings 1974. Edited by H. Frisch. VI, 356 pages. 1976.

Vol. 120: G. Hasenkamp, Specification and Estimation of Multiple-Output Production Functions. VII, 151 pages. 1976.

Vol. 121: J. W. Cohen, On Regenerative Processes in Queueing Theory. IX, 93 pages. 1976.

Vol. 122: M. S. Bazaraa, and C. M. Shetty, Foundations of Optimization VI. 193 pages. 1976

Vol. 123: Multiple Criteria Decision Making. Kyoto 1975. Edited by M. Zeleny. XXVII, 345 pages. 1976.

Vol. 124: M. J. Todd. The Computation of Fixed Points and Applications. VII, 129 pages. 1976.

Vol. 125: Karl C. Mosler. Optimale Transportnetze. Zur Bestimmung ihres kostengünstigsten Standorts bei gegebener Nachfrage. VI, 142 Seiten. 1976.

Vol. 126: Energy, Regional Science and Public Policy. Energy and Environment I. Proceedings 1975. Edited by M. Chatterji and P. Van Rompuy. VIII, 316 pages. 1976.

Vol. 127: Environment, Regional Science and Interregional Modeling. Energy and Environment II. Proceedings 1975. Edited by M. Chatterji and P. Van Rompuy. IX, 211 pages. 1976.

Vol. 128: Integer Programming and Related Areas. A Classified Bibliography. Edited by C. Kastning. XII, 495 pages. 1976.

Vol. 129: H.-J. Lüthi, Komplementaritäts- und Fixpunktalgorithmen in der mathematischen Programmierung. Spieltheorie und Ökonomie. VII, 145 Seiten. 1976.

Vol. 130: Multiple Criteria Decision Making, Jouy-en-Josas, France. Proceedings 1975. Edited by H. Thiriez and S. Zionts. VI, 409 pages. 1976.

Vol. 131: Mathematical Systems Theory. Proceedings 1975. Edited by G. Marchesini and S. K. Mitter. X, 408 pages. 1976.

Vol. 132: U. H. Funke, Mathematical Models in Marketing. A Collection of Abstracts. XX, 514 pages. 1976.

Vol. 133: Warsaw Fall Seminars in Mathematical Economics 1975. Edited by M. W. Loś, J. Loś, and A. Wieczorek. V. 159 pages. 1976.

Vol. 134: Computing Methods in Applied Sciences and Engineering. Proceedings 1975. VIII, 390 pages. 1976.

Vol. 135: H. Haga, A Disequilibrium – Equilibrium Model with Money and Bonds. A Keynesian – Walrasian Synthesis. VI, 119 pages. 1976.

Vol. 136: E. Kofler und G. Menges, Entscheidungen bei unvollständiger Information. XII, 357 Seiten. 1976.

Vol. 137: R. Wets, Grundlagen Konvexer Optimierung. VI, 146 Seiten. 1976.

Vol. 138: K. Okuguchi, Expectations and Stability in Oligopoly Models. VI, 103 pages. 1976.

Vol. 139: Production Theory and Its Applications. Proceedings. Edited by H. Albach and G. Bergendahl. VIII, 193 pages. 1977.

Vol. 140: W. Eichhorn and J. Voeller, Theory of the Price Index. Fisher's Test Approach and Generalizations. VII, 95 pages. 1976.

Vol. 141: Mathematical Economics and Game Theory. Essays in Honor of Oskar Morgenstern. Edited by R. Henn and O. Moeschlin. XIV, 703 pages. 1977.

Vol. 142: J. S. Lane, On Optimal Population Paths. V, 123 pages. 1977.

Vol. 143: B. Näslund, An Analysis of Economic Size Distributions. XV, 100 pages. 1977.

Vol. 144: Convex Analysis and Its Applications. Proceedings 1976. Edited by A. Auslender. VI, 219 pages. 1977.

Vol. 145: J. Rosenmüller, Extreme Games and Their Solutions. IV, 126 pages. 1977.

Vol. 146: In Search of Economic Indicators. Edited by W. H. Strigel. XVI, 198 pages. 1977.

Vol. 147: Resource Allocation and Division of Space. Proceedings. Edited by T. Fujii and R. Sato. VIII, 184 pages. 1977.

Vol. 148: C. E. Mandl, Simulationstechnik und Simulationsmodelle in den Sozial- und Wirtschaftswissenschaften. IX, 173 Seiten. 1977.

Vol. 149: Stationäre und schrumpfende Bevölkerungen: Demographisches Null- und Negativwachstum in Österreich. Herausgegeben von G. Feichtinger. VI, 262 Seiten. 1977.

Vol. 150: Bauer et al., Supercritical Wing Sections III. VI, 179 pages. 1977.

Vol. 151: C. A. Schneeweiß, Inventory-Production Theory. VI, 116 pages. 1977.

Vol. 152: Kirsch et al., Notwendige Optimalitätsbedingungen und ihre Anwendung. VI, 157 Seiten. 1978.

Vol. 153: Kombinatorische Entscheidungsprobleme: Methoden und Anwendungen. Herausgegeben von T. M. Liebling und M. Rössler. VIII, 206 Seiten. 1978.

Vol. 154: Problems and Instruments of Business Cycle Analysis. Proceedings 1977. Edited by W. H. Strigel. VI, 442 pages. 1978.

Vol. 155: Multiple Criteria Problem Solving. Proceedings 1977. Edited by S. Zionts. VIII, 567 pages. 1978.

Vol. 156: B. Näslund and B. Sellstedt, Neo-Ricardian Theory. With Applications to Some Current Economic Problems. VI, 165 pages. 1978.

Vol. 157: Optimization and Operations Research. Proceedings 1977. Edited by R. Henn, B. Korte, and W. Oettli. VI, 270 pages. 1978.

Vol. 158: L. J. Cherene, Set Valued Dynamical Systems and Economic Flow. VIII, 83 pages. 1978.

Vol. 159: Some Aspects of the Foundations of General Equilibrium Theory: The Posthumous Papers of Peter J. Kalman. Edited by J. Green. VI, 167 pages. 1978.

Vol. 160: Integer Programming and Related Areas. A Classified Bibliography. Edited by D. Hausmann. XIV, 314 pages. 1978.

Vol. 161: M. J. Beckmann, Rank in Organizations. VIII, 164 pages. 1978.

Vol. 162: Recent Developments in Variable Structure Systems, Economics and Biology. Proceedings 1977. Edited by R. R. Mohler and A. Ruberti. VI, 326 pages. 1978.

Vol. 163: G. Fandel, Optimale Entscheidungen in Organisationen. VI, 143 Seiten. 1979.

Vol. 164: C. L. Hwang and A. S. M. Masud, Multiple Objective Decision Making – Methods and Applications. A State-of-the-Art Survey. XII, 351 pages. 1979.

Vol. 165: A. Maravall, Identification in Dynamic Shock-Error Models. VIII, 158 pages. 1979.

Vol. 166: R. Cuninghame-Green, Minimax Algebra. XI, 258 pages. 1979.

Vol. 167: M. Faber, Introduction to Modern Austrian Capital Theory. X, 196 pages. 1979.

Vol. 168: Convex Analysis and Mathematical Economics. Proceedings 1978. Edited by J. Kriens. V, 136 pages. 1979.

Vol. 169: A. Rapoport et al., Coalition Formation by Sophisticated Players. VII, 170 pages. 1979.

Vol. 170: A. E. Roth, Axiomatic Models of Bargaining. V, 121 pages. 1979.

Vol. 171: G. F. Newell, Approximate Behavior of Tandem Queues. XI, 410 pages. 1979.

Vol. 172: K. Neumann and U. Steinhardt, GERT Networks and the Time-Oriented Evaluation of Projects. 268 pages. 1979.

Vol. 173: S. Erlander, Optimal Spatial Interaction and the Gravity Model. VII, 107 pages. 1980.

Vol. 174: Extremal Methods and Systems Analysis. Edited by A. V. Fiacco and K. O. Kortanek. XI, 545 pages. 1980.

Vol. 175: S. K. Srinivasan and R. Subramanian, Probabilistic Analysis of Redundant Systems. VII, 356 pages. 1980.

Vol. 176: R. Färe, Laws of Diminishing Returns. VIII, 97 pages. 1980.

Vol. 177: Multiple Criteria Decision Making-Theory and Application. Proceedings, 1979. Edited by G. Fandel and T. Gal. XVI, 570 pages. 1980.

Vol. 178: M. N. Bhattacharyya, Comparison of Box-Jenkins and Bonn Monetary Model Prediction Performance. VII, 146 pages. 1980.

Vol. 179: Recent Results in Stochastic Programming. Proceedings, 1979. Edited by P. Kall and A. Prékopa. IX, 237 pages. 1980.

Vol. 180: J. F. Brotchie, J. W. Dickey and R. Sharpe, TOPAZ – General Planning Technique and its Applications at the Regional, Urban, and Facility Planning Levels. VII, 356 pages. 1980.

Vol. 181: H. D. Sherali and C. M. Shetty, Optimization with Disjunctive Constraints. VIII, 156 pages. 1980.

Vol. 182: J. Wolters, Stochastic Dynamic Properties of Linear Econometric Models. VIII, 154 pages. 1980.

Vol. 183: K. Schittkowski, Nonlinear Programming Codes. VIII, 242 pages. 1980.

Vol. 184: R. E. Burkard and U. Derigs, Assignment and Matching Problems: Solution Methods with FORTRAN-Programs. VIII, 148 pages. 1980.

Vol. 185: C. C. von Weizsäcker, Barriers to Entry. VI, 220 pages. 1980.

Vol. 186: Ch.-L. Hwang and K. Yoon, Multiple Attribute Decision Making – Methods and Applications. A State-of-the-Art-Survey. XI, 259 pages. 1981.

Vol. 187: W. Hock, K. Schittkowski, Test Examples for Nonlinear Programming Codes. V. 178 pages. 1981.

Vol. 188: D. Bös, Economic Theory of Public Enterprise. VII, 142 pages. 1981.

Vol. 189: A. P. Lüthi, Messung wirtschaftlicher Ungleichheit. IX, 287 pages. 1981.

Vol. 190: J. N. Morse, Organizations: Multiple Agents with Multiple Criteria. Proceedings, 1980. VI, 509 pages. 1981.

Vol. 191: H. R. Sneessens, Theory and Estimation of Macroeconomic Rationing Models. VII, 138 pages. 1981.

Vol. 192: H. J. Bierens: Robust Methods and Asymptotic Theory in Nonlinear Econometrics. IX, 198 pages. 1981.

Ökonometrie und Unternehmensforschung
Econometrics and Operations Research